THE
TEACHING
BRAIN

THE
TEACHING
BRAIN

AN EVOLUTIONARY TRAIT
AT THE HEART OF EDUCATION

VANESSA RODRIGUEZ
WITH MICHELLE FITZPATRICK

THE NEW PRESS

NEW YORK
LONDON

Requests for permission to reproduce selections from this book should be mailed to: Permissions Department, The New Press, 120 Wall Street, 31st floor, New York, NY 10005.

Published in the United States by The New Press, New York, 2014
Distributed by Perseus Distribution

ISBN 978-1-59558-996-5 (hardcover)
ISBN 978-1-62097-022-5 (e-book)
CIP data available.

The New Press publishes books that promote and enrich public discussion and understanding of the issues vital to our democracy and to a more equitable world. These books are made possible by the enthusiasm of our readers; the support of a committed group of donors, large and small; the collaboration of our many partners in the independent media and the not-for-profit sector; booksellers, who often hand-sell New Press books; librarians; and above all by our authors.

www.thenewpress.com

Book design and composition by Bookbright Media
Illustrations by Teresa Peitrowski and Michael Peitrowski
This book was set in Goudy Oldstyle and Futura

Printed in the United States of America

10 9 8 7 6 5 4 3 2 1

"It would be very strange for people to see someone without [a daemon]. It'd be just as strange as seeing someone without a head. Someone without a daemon would be considered horribly mutated—missing something essential."

—*Chris Weitz, director of* The Golden Compass

This book is dedicated to my daemon—without him none of this would be possible.

"We have a lot of rocks in the classrooms . . . we study rocks. We're learning about prehistory. We're learning about the beginnings of the planet. We're learning about patterns and cycles. [A visitor to our school] once said, 'Why are there so many rocks in the school? What can you do with a rock? You can write a story about a rock. You can learn a little about rocks—then you're done.' And I thought, no. You can spend a lifetime with a rock. And that is the way I feel about teaching anything."

—*"Liz," a fourth-grade teacher in New York*

This book is also dedicated to all the teachers who believe you can spend a lifetime with a rock.

CONTENTS

INTRODUCTION

There's no shortage of complaints about the state of schooling today, no shortage of calls to reform our educational system, no shortage of opinions on how to fix what many (even those who fervently disagree on the particulars) agree amounts to a major social dilemma. The contemporary challenges facing formal education in the United States are vast, complicated, and emotionally charged. Educators, scientists, psychologists, government officials, and bestselling authors are all part of the mix of voices creating the conversation and, in part, prolonging the controversies, both warranted and not.

But it seems the biggest elephant in the proverbial room is how we think about teaching and our teachers. All sides of the teaching debate have sought to define "good teaching." However, this effort is as misguided as one that would label a student a "good learner." We need to keep the terms "good" and "teaching" forevermore apart. Indeed, the concept of a perfect teacher for all students is a complete myth. Instead we need to be asking new questions.

Our questions should begin with one in particular: What is teaching?

A five-year-old child, absent any formal training of the sort classroom teachers get, easily "teaches" a younger child how to construct a tower of blocks, could begin to read her two-year-old brother a story from her favorite book, or could show her friend how to make a tea party—although half of that party may be imaginary. As children grow, these social interactions develop, becoming multilayered and more complicated. A ten-year-old can show a teammate how to kick a soccer ball properly. A teenager might help his buddy understand the difference between adding and subtracting positive and negative numbers. To do that he might reach for a pencil and paper to write out an example, or go online to find a sample problem on a math website. Adding organization and pre-thought to teaching moments reflects a more complex form of thinking.

Or consider what parents do quite instinctively with their young children. The earliest forms of learning for any baby begin with the parent relationship. A mother talks to a toddler about what they see, hear, and sense around them: "Look at the birdie in the tree!" "Smell the yummy toast with butter." "Take my hand, and let's walk over to feed the ducks." Parents, without thinking, are constantly teaching their children about the world around them, talking to them so that they begin to develop their vocabulary, and interacting with them so that children begin to understand the simplest concepts of sequence, procedural knowledge such as tying shoes and using a spoon, and more concrete knowledge about colors, numbers, or the sight and sound of letters.

So what, precisely, is teaching? This book answers that question in a new, transformative, and perhaps surprising way: teaching is a human, evolutionary skill. Although we may not all be in a classroom, we are all teachers.

Like learning, teaching is a skill (or set of skills) that develops over time. Social interactions of the sort mentioned above happen naturally all the time, yet most of us do not define them as the teaching and learning moments that they are. Every social interaction does not involve teaching, but—unlike moments in life when we are learning—every teaching moment is, in fact, a social interaction.

As much as the five-year-old, the ten-year-old, the teenager, and the parent are all teachers, we know instinctively, and through research, that each of them teaches differently. Certainly their age and varied cognitive abilities contribute to the difference, as do a host of other factors that are explicitly and implicitly connected to the social nature of teaching. For example, if all of these teachers set about the task of showing another how to make a tea party, each of those interactions would look markedly different based on who is doing the teaching and who is doing the learning. The key, however, is that both actors—teacher *and* learner—are equally important to the transfer, understanding, and creation of knowledge and skills. Here is where things start to get sticky.

As the process and development of teaching become professionalized in more formal settings, suddenly the entire premise of the interaction ceases to be a natural interplay between the teacher and the learner. Inside school walls and pedagogical framing, teaching is often understood and practiced as if it were a one-way street, with a mostly unidirectional focus on the learner—the student. Who the teacher is and how she is processing these same moments become secondary, and in some cases irrelevant, to the needs and process of the student. And this is to say nothing of how any benefit to the teacher—either professional or personal—is almost always downplayed, if not overlooked entirely. This is why even thoughtful, well-intentioned efforts and policies that aim exclusively to be "student-centered" are incomplete. The process of teaching and learning is an inherently dynamic and social one, yet we press on in schools and in life with a harmfully imbalanced perspective. Everyone suffers when only the interests of the student seem to matter. Half of the equation is lost.

HOW I CAME TO WRITE THIS BOOK

I spent more than a decade teaching in the classroom—science, history, and English. Truth be told, I was a combative teacher. I loved my students, taught to their individual needs, and was devoted to helping them learn. However, I was not good at being part of a system that often seemed at odds with their learning, and with my own teaching process. My methods often came into conflict with what the school's goals—or more precisely, standards—expected of me. There was a lack of fit.

Most of the time, I resisted allowing this lack of fit to define my approach. Like many teachers I know, I taught because I found it fascinating and rewarding, especially when I was in sync with my students. The principals I worked for would often say, "But in the end, Vanessa, you're doing it for the children, right?" My answer was often silence. Dare I ever admit that I was a professional teacher because I found teaching so personally fulfilling? I taught because the act of teaching spurred growth within me; it meant something to me, and contributed to a more intelligent, informed, and connected world. It was not *just* about the children I was helping.

I know with confidence (though slight discomfort) that I never would have chosen to teach if it were purely an altruistic act. I certainly wouldn't have taught for so long if that were the case. The moment I realized that my teaching methods and philosophy were no longer in line with major education reform efforts across the city where I was teaching, I knew it was time for me to move on. The balance had changed dramatically, and my vision for teaching no longer fit—even awkwardly—in the school system I had been so committed to teaching in.

Still, I knew I had been a successful teacher. My students exhibited clear development in their learning—not just in measures that mattered little to me, such as standardized exams, but in quality measures such as their writing ability, sophisticated communication, debate skills, and, most important, their sense of identity. I was proud to facilitate and support their development. While each day was a challenge, it was also extremely rewarding to see students develop and even more so to form personal bonds with them. Their development was inextricably linked to my development as a teacher.

I believe the reason many teachers leave the profession is not some amorphous burnout. I suspect (and to a modest extent, I know) that many teachers who move on do so for a similar reason: their perspective on teaching and that of the system within which they teach do not jibe. Even after years of trial and error, reaching synchrony with an administration or system—the kind of synchrony that effective, successful teachers find with their students—can still seem like a distant dream. Furthermore, the various other reasons teachers report feeling undervalued do nothing to help the situation.[1]

My own experiences teaching, the ups and the downs, prompted me to wonder about the drive to help others learn, the urge and the practice commonly referred to as teaching. What do we really mean by teaching? Do we all have the same thing in mind when we conceptualize teaching? Who should we trust to say what teaching is, and how does this play out in schools? How can we even begin to transform education policy or practice unless we all understand what teaching really means?

Four years ago, in the midst of my graduate-school research, I made a startling—and, it turns out, profound—connection between the cognitive psychology and neuroscience I had been studying and the practice of teaching: for all we know about the nature and science of learning, especially the discoveries in brain research, we have grown very little in our insight into the teaching process. I came to understand that we all do it, this act we call teaching. But why do humans teach? How do we teach? And why has teaching, an interaction so integral to the foundation of education, been given such short shrift? Quite simply it is because, despite mountains of books and research on pedagogy (the practice of teaching), no one has ever truly bothered to understand specifically how the teaching process and its corollary, the teaching brain, are *separate and distinct* from the learning process and the learning brain.

This book illuminates how both past and current definitions of teaching are outdated and don't match the vast research in the learning sciences and the more fledgling area of neuroscience. Looking beyond the lab, commonly accepted definitions of teaching don't even match the actual practice of human teaching. It's time for that to change.

Starting with the premise that learning and teaching, while inextricably related, are indeed distinct processes, we can begin to develop a clearer picture of the teacher in all of us, whether we are in a classroom or a boardroom. New doors open up and everyone is better served when we demystify teaching based on a complete understanding of the cognitive, biological, and psychological processes of the brain. Through piecing together this puzzle, my goal is to unearth a new definition of teaching that is at once revolutionary and profoundly commonsense, one that builds on our understanding of learning yet has unique, individual properties. Part of a larger thrust in the learning sciences to explore and understand the human brain, the research that supports the model of the teaching brain connects the dots between what we have come to understand about the learning brain and how that relates to an analogous teaching brain. We know that learning is much more than memorizing content. It's about thinking skills and the ability to transfer knowledge to an array of situations. We don't expect learners to be expert learners right away; we know that becoming a resourceful, self-regulated, persistent learner happens over time. Learners develop. The same is true of teachers.

THE STRUCTURE, CONTEXT, AND PURPOSE OF *THE TEACHING BRAIN*

The pages that follow aim to shift how we see and understand teaching. This book is also designed to serve as a useful tool for teachers of all types, and will be particularly helpful to professional teachers looking to improve their skills. Part One starts by debunking older models of teaching in which knowledge is like water—something a teacher can simply pour into the mind of a student. This section also highlights misguided theories and policy reforms that wield a hefty amount of influence on both lay and professional perceptions of teaching. Beginning with an examination of the science behind the learning brain, Part Two lays out a revolutionary new framework for understanding the teaching brain as a dynamic, interactive system created by both teacher and student. The book then finishes with an exploration of how this new framework can help teachers and education reformers become more effective and successful, and includes a discussion of exciting neurobehavioral research that is using cutting edge imaging technology to observe interpersonal brain interactions (e.g., teaching) in real time.

Different from other brain-based educational advice that touts ways to train your brain or follow simple "rules" to enhance productivity and learning, *The Teaching Brain* is based on a carefully honed, research-driven understanding of teaching as a human-specific, natural ability, and a building block of the social fabric of humanity. There are those who will likely raise an eyebrow at the assertion that teaching is specific to humans. Yes, an ant can teach another ant how to find food, monkeys can teach their young how to use a stick to catch their prey, and cats and dogs pass on survival skills as well, but let's take a step back and ask ourselves whether that is the extent of the type of teaching that we do as humans. I doubt it is the type of teaching that we anticipate our children will get in school. There's plenty of room for debate on whether animals teach or whether teaching is human-specific. However, that argument, and any proposed resolutions to it, will not bring us any closer to understanding, in real and practical terms, what humans do when we teach. Regardless of whether and how human teaching is more sophisticated than that of animals, discovering the processes of this complex cognitive skill in humans is essential.

Unfortunately, the current definitions and interpretations of teaching are inadequate; they are tied only to our understanding of learning. Until we clearly define and understand teaching, our efforts to reform education are doomed to fail. More important, once we recognize and embrace the teaching brain in all of us, we will be better equipped to comprehend the learning challenges that we all face in our homes, schools, and places of work. This book will enable any of us to find the teacher within, the teacher who knows best how to help others in our lives.

It may seem strange that we don't really understand teaching, since most of us probably think that we know it when we see it and that this is good enough. In some sense we believe that teaching is simply a service, a duty to children that, like parenting, therefore requires no further definition. If we consider the role of teachers the way we consider the role of parents, teaching is cast as an obligation that cannot be dissected, and teachers simply must give everything they can. This mode of thinking helps neither teachers nor students. As modern psychology and other behavioral research shows, there are indeed more and less effective modes of parenting.[2] There are endless approaches that can be tailored to the nuances of each individual family situation. The time has come for a similar evolution in our thinking about teaching.

Problems abound in today's education system—from dismal graduation rates and a lack of school financing to "bad" teachers and "underperforming" kids. Reformers of every sort have matched solutions to these problems: more after-school programming, more money, more testing of kids, and better teachers. Yet if we don't see any problem with our definition of teaching, we are missing out on the most important solutions. Without accurately defining teaching, how it works, its processes, and its systems, we don't have a clear understanding of the cornerstone of our education system—a system that so many work tirelessly to improve. Just as we cannot suggest how to build a better parent, it is not possible to simply outline how we can "build a better teacher."[3] We can build machines, but we can't build humans. To improve human behavior, we must understand it. Observation is crucial, but we come to understand behavior on a deeper level through studying the brain. The key to unlocking the mystery of teaching is to understand the teaching brain.

THE
TEACHING
BRAIN

PART ONE

WHAT IS WRONG WITH OUR DEFINITION OF TEACHING?

1

THE THEORIES THAT HAVE LED US ASTRAY

My first teaching position was in a very affluent neighborhood in the suburbs outside New York City. Typically teachers roughed it out for years in urban public schools, trying to gain enough credentials to land such a position. For me this was the result of a mistake. I hadn't realized that, after graduating with a teaching degree, I should have gone into the Board of Education pool with other new teachers seeking jobs within New York City. Had I done so, I would have been "matched" and placed in a vacant position at a random public school. Instead I scoured through the education section of the *New York Times* every Wednesday and applied to every English position I found listed.

So there I was, standing in my brand-new classroom with a crisp, clean copy of the new teacher manual. During the first English department meeting I attended, I was given a tried-and-true curriculum guide, compiled by the department head over years of teaching high school students. All I had to do, I was told, was choose from an approved list the books that I would read with my class, then pull those books from the department storage room.

Upon entering the storage room, I was greeted with rows and rows of shiny new books. I quickly envisioned myself standing in the front of my classroom, book in hand, reciting line after line from Shakespeare as students absorbed every word. I would be like a conductor bringing forth and organizing the beautiful music created by past masters. The results would be magical: as my audience, the students, heard this masterpiece, their minds would travel to previously unknown worlds for them to explore. I wanted nothing more than to share with students the way I understood Shakespeare. I was certain that I could get them to love the works by seeing them the way that I did. Surely, I thought, their apathy was the result of disinterested grown-ups. I wanted them to really learn, as opposed to just regurgitating facts.

The senior teachers cautioned me against using too much passion in my classroom: "Just read the books, ask them questions of comprehension, and have them write essays to prove that they've learned the appropriate theme of the book." If I could do that while also convincing students to love the literature, I thought, then I was being a great teacher. After all, I was the one holding all the knowledge, right? I just needed to ignite my students' passions to free them of the shackles affixed by past teachers and experiences. My goal was to give all of my knowledge to my not-yet-hungry students. I was sure the key to success was simply to motivate them to a point where they were so excited to receive the knowledge that they would ask for it—thirst for it, even.

I often try to make myself feel better by remembering that this definition of teaching has been around for at least fifty years. These erroneous assumptions of best teaching practices have led us astray when we train our teachers, establish expectations, and design schools, testing, and curricula. To be blunt, our current models of teaching are outdated and unsophisticated. Their deterministic and rigid criteria stand in the way of fully integrating current research from the learning sciences and what we know of how the brain works.

As counterintuitive as it may seem, there is quite a bit of evidence that we have structured our system of education on animal behavior experiments, utilizing definitions of how animals teach.

PAVLOV, PIED BABBLERS, AND PUNISHMENT

One of the most common assumptions about teaching comes out of behaviorist research on learning. The behaviorist view of learning is one in which learners are seen as empty vessels, ready to be filled with knowledge. In their view learning is an additive process where information is *held* by experts and *given* to naive learners who are otherwise a blank slate. Pavlov is perhaps most notable for his contribution to the foundation upon which behaviorist learning theories are built. His studies detailed how animals react to stimuli in response to the environment around them. We all know of the classic experiment involving Pavlov's dogs, who learned to expect food every time they heard the sound of a bell ring. Once the dogs learned that the bell meant food, Pavlov could watch the dogs salivate each time he rang the bell.

Psychologist John B. Watson saw great potential in Pavlov's design and decided that this method of stimulus response could tell him quite a bit about human learning. He believed that people learned due to continued experience with different stimuli in their environment. Learning was considered a conditioned response, and in his view anyone could be conditioned to have a certain response. Once a behavior was learned, it could be adapted based on changes in the environment. Watson believed that understanding the range of human behaviors was the key to understanding learning.

To be clear, it's quite possible for a person to be trained to re-act a certain way. Even if you know you'll be sorely disappointed with the outcome, you might still feel compelled to buy a Subway sandwich when you get a whiff of that bread aroma while walking down the street. This reaction is not mediated by logic flowing through your mind. It's just a reaction that you've been trained to follow—your free will basically takes a step back.

B.F. Skinner drew on this concept to design a system that he felt would guarantee preferred behavior. He believed that one could use rewards and punishment to influence behavior and learning, and his model of operant conditioning claimed that the mind, or one's thoughts and beliefs, were not relevant factors in decision making. All processes, he argued, were the result of either a reward or punishment received at the end of a task, not the product of free will.

These concepts were an important influence in the way behaviorists viewed teaching. Their definition of teaching was based on what could be observed. In the behaviorist view, teaching is an act where the teacher changes his or her behavior to aid a naive student in acquiring knowledge or skills.[1] Let's take a common example from the animal kingdom.

Pied babblers are birds who use special calls to lure their young away from danger. The training begins young, when the birds are first learning to eat. The mother bird makes this special call, and each time she calls, she then feeds her baby.[2] It's very Pavlovian: over time the young bird learns that when he hears the call, he gets food. When they are old enough to leave the nest, the mother bird can use this special call to keep her young away from dangerous situations. Granted, she ends up expending a lot of energy and food to solidify these life lessons, but the outcome can be lifesaving.

This definition of teaching makes perfect sense in conjunction with the behaviorist view of learning. Throughout the interaction the teacher is providing the student with feedback, and as a result, the students have learned the knowledge or skill more effectively than they would have had they not had the interaction. *Note that this type of interaction yields no direct benefit to the teacher.*[3] In fact, the mother bird could very well go hungry for a few days while her offspring are first learning the call. However, think of the benefit that does exist. This is how humans and animals learn to catch food, locate watering holes, and in general survive. This logic, if oversimplified or misapplied, could easily lead to the conclusion that being a selfless teacher is just something we all have to do. This bird example is actually quite popular among biologists, who argue that animals, just like humans, can teach and that they do so in the same way as we humans. Some researchers argue that teaching in fact evolved independently of learning and is therefore able to be observed and quantitatively tested.[4]

In this frame, teaching looks something like this: the mother bird and baby bird are cooperating, and we can see from their behavior that the payoff is dependent on the response. That is, in order to receive food, the baby must come running (or flying) when it hears the special call, otherwise it will starve. And the mother must issue the appropriate call when there is food to be had. The whole interaction functions purely to facilitate learning in the young bird. The donor (the mother bird, who has absolute knowledge and power) gives information to the receiver. The mother bird gathers the food and chooses when and where to release it. The sole beneficiary of the teaching is the baby bird, for presumably this is the only way the baby will learn how to eat and to be part of the pied babbler community. This is how adults within a community teach the young both to survive and to participate in the culture. It's a necessary part of building organized societies.

A CLOSER LOOK AT BEHAVIORIST VIEWS OF TEACHING

Skinner's suggestions for improving teaching followed this animal model. Skinner believed that humans and animals were cut from the same cloth. He thought that teaching could be completed utilizing a stimulus response, meaning that teachers could input or transmit information to students, not unlike pouring juice into a glass or filling up a bowl with rice, and the students would learn the information, supposedly because they were being given positive rewards to do so. To prove his theory, Skinner designed a teaching machine, which modeled how he thought the interaction should occur.[5] The machine was programmed to give students a list of questions to respond to and to reward correct answers. The machine provided students with immediate personalized feedback and new questions based on their responses, led students through correcting their errors, and automatically adjusted to the pace and level of assistance required by the learner.[6] Skinner believed this individualized instruction guaranteed that students would pay attention and remain motivated. Teaching machines were thought to be useful in teaching actual content, self-management, and decision making.[7] In the ensuing decades, computer-assisted instruction such as this became a driving force in the programmed-instruction movement.[8] Companies such as IBM partnered with universities including Stanford to design computer-assisted instructional systems to teach an entire class of students.[9]

The history of the behaviorist view of learning matters because it developed into a widely accepted view of teaching that, along with other threads in the history of learning and teaching, still hold considerable sway over how we make educational decisions today, even as our understanding of learning has undergone a profound shift over time, from a unidimensional view of a learner being filled with knowledge to a recognition of the learner's brain as a dynamic, context-dependent system.[10] The transmission model of learning takes many forms. It is most easily viewed by looking at behaviorist frameworks, most significant among them the work of Pavlov, Watson, and Skinner. In a nutshell, these behaviorists argued that learning was the product of either positive or negative reinforcement of a response to a stimulus. Skinner took the concept of tabula rasa one step further to create the doctrine of the empty organism, which postulates that a person is like "a vessel to be filled by carefully designed experiences."[11]

In contrast to the behaviorists, cognitive theorists viewed learning as the acquisition or reorganization of mental structures through which humans process and store information. Piaget was the first to systematically study and develop principles of human cognitive development and learning. Founded on these principles was the belief that learners construct knowledge from their experiences.[12] Adding to this foundation, Vygotsky's theory of social cognition further argued that learners develop while interacting with their world. Utilizing these principles as well as those of the constructivists, including Dewey, Montessori, and Kolb, who believed that children learn most effectively when they are given opportunities to co-create knowledge, Kurt Fischer and others drew inspiration from dynamic systems and extended cognitive theories by developing a flexible model of how the learning brain develops from birth through adulthood.[13] This theory posited that the learner's brain is a complex, dynamic system. Therefore the development of learning does not follow a linear pattern. Instead, there are learning spurts that are due to shifts in context and level of support.

This concept of cognitive development is called dynamic skill theory (which you will read more about in chapter 3) and is increasingly being substantiated on a neurologic level by new studies of the learning brain that use brain imaging and other technologies to identify the neural processes and pathways through which learning develops over time. This integrated approach of complex, dynamic cognitive theoretical frameworks and neurologic mapping of learning pathways is the foundation of the popular modern view of brain-based learning.[14]

Still, behaviorist principles are alive and well in our everyday educational lives—they drive state-mandated exams, charter schools that rely on "no excuse" cultures, incentivized teaching models, and several other current education reform policies. While the next chapter will go into greater detail on how this model plays out in classrooms today, let's review quickly what behaviorist theory, drawing primarily from animal behavior studies, sees as the three main aspects of teaching:[15]

1. **A cooperative donor-receiver behavior.** The teacher purposely alters his or her behavior in the presence of the student in order to facilitate learning.

2. **Teaching is a selfless act.** It comes at a cost to the teacher, who does not directly benefit from the interaction.

3. **The interaction allows a student to learn efficiently.** A student learning on his or her own would learn less efficiently.

There are a number of problematic repercussions to the underlying assumptions inherent in these three characteristics of behaviorist teaching. This definition of teaching is firmly rooted in animal studies on learning, and no proven, obvious, or sensible reason exists for applying it to humans, let alone for using it to design entire school systems. And humans' ability to read and respond to the body language of others, which is something that takes place in a teaching interaction, is beyond the abilities of other animals that have been seen to teach.

Even though the behaviorist model of input-output comes up short in describing the broad range of the teaching skill, that's not to say that it is entirely wrong. I am not so bold as to fly in the face of a wealth of significant research in this area. The point is that these Skinner-esque models of teaching are a mere sliver of the range of teaching that is capable by humans. The cognitive processing required to produce the teaching output we see in behaviorist models is quite low.

Yet many researchers, economists, and policy makers continue their attachment to this basic definition, which makes teaching into a largely functional endeavor. They do so primarily because it provides testable criteria by which to measure learning and therefore (by their estimation) teaching. While this input-output model offers a clean and linear framework to understand and assess teaching and learning, it is harmfully oversimplified.

THE PROBLEM WITH THE EMPTY-VESSEL THEORY OF LEARNING

Indeed, there is a problem with treating learners as mere empty vessels. Pouring knowledge into the passively receptive head of a student leaves the student to conceive of knowledge as something that already exists, rather than something that is created or co-constructed. In this scenario, the teacher is little more than a learning tool, transferring knowledge from one vessel (the teacher) to another (the student). This is precisely why Skinner believed that a teaching machine could do the job of a classroom teacher. The machine was programmed to enact the behavior of a knowledge expert transmitting information to the naive student and offering feedback in the form of rewards and punishment on the student's answers, voilà!—a ready-made teacher. Recall that Skinner was one of the first to suggest that a teacher could be something other than a human. In his model and that of other behaviorist believers, a teacher could be a television, a computer program, or even an ant. A teacher is anything that *causes* you to learn. This is of course a summarized explanation, but that's the gist.

The behaviorists are not the only researchers who hold an empty-vessel view of learning. Over the years, many pedagogical approaches—that is, theories of teaching—have been premised on a similar view: that novice learners obtain knowledge through interaction with their expert teachers. If teachers are merely learning tools, then it follows that they too just need to be filled with any information necessary for the learner to acquire knowledge. It is this definition that ignited the current excitement for new modes of technology-based "teaching" (via the Internet, computers, and TV). Learning tools that are programmable, and ostensibly more reliable because of their lack of human error, are in many cases cheaper knowledge transmitters. Initially that might seem like a great relief—we don't need more money or more teachers to improve schools; we just need machines. Unfortunately, any definition of teaching that is premised on learners being empty vessels and teachers being mere transmitters of knowledge and information fails to take account of the latest discoveries in the learning sciences.

In stark contrast to the empty-vessel theory, it turns out that learning is a dynamic and complex behavior (more to come on this in chapter 3). When we unravel the complexity of learning, we see why so many well-intentioned approaches to supporting teaching and learning have not worked. As we work to build this new definition of teaching, it's useful to take a look at some other commonly held misconceptions about teaching and learning.

THEORY OF MIND: TEACHING AS INNATELY HUMAN

More recently, another attempt to define teaching emerged from cognitive psychology, best exemplified by the work of David Premack and Sidney Strauss, who noted that teachers, like all humans, demonstrate a theory of mind (ToM). *Theory of mind* refers to a person's ability to understand what is going on in the mind of another human. Young children have an emerging ToM: at one and two years old, they mostly believe that the world revolves around them; the thoughts or emotions of another are not experienced as separate from themselves. Around the ages of three and four years, children begin to show an awareness that others do in fact have thoughts or feelings different from their own. This growing awareness represents important developmental milestones—social, emotional, and cognitive.

Adults utilize ToM all the time. Theory of mind is what we all use when we are trying to understand what makes another person tick, so that we know how to engage and react. It's what we might do on a first date in trying to figure out how to impress the person we're with and get a second date. We do it when we're negotiating with a salesclerk, hoping to get a better deal on an item. You are even doing it right now as you read this book: you are trying to follow my train of thought as I introduce you to the concept of the teaching brain.

Premack has defined ToM as a necessary characteristic of human teaching.[16] Developmental psychologists assert that ToM enables teachers to take into account the needs of the learner as they plan, evaluate, and reorganize their teaching.[17] In his seminal work Strauss notes that teaching is a natural human cognitive ability because only humans have ToM.[18] Animals do not have ToM; they do not modify their teaching based on the learner's ability and progress.[19] Teaching among animals involves simple responses to behavioral cues; animals do not adapt to individual variation.[20]

Allen Pearson has described teaching among humans as the teacher's intention to help the learner attain knowledge.[21] This human-specific intention to teach goes well beyond survival instinct; it has as goals uniting individuals and closing the "knowledge gap" between them.[22] In order to achieve these goals, humans adapt their teaching based on learner variation. In this perspective, understanding how teachers think is necessary if we are to grasp an understanding of the human cognitive skill of teaching.

At first glance, ToM seems to present a revolutionary new way of understanding teaching, casting it as an innate intellectual and psychological skill. Indeed, the constructivists have utilized ToM to develop and design curricula that emphasize co-creation of knowledge between learner and teacher. And while ToM gets us one step closer—it offers valuable insight into teaching as a process—it too is incomplete as an explanation of teaching as an interaction.

Let's start with what we know about infants and teaching. In a study of one-year-old nonverbal children, researchers found evidence that they freely gave information to aid the experimenter.[23] The experimenter intentionally pushed an object off the table in plain sight and then feigned looking for it before telling the child that he or she had lost the object and asking whether the child had seen it. The child offered up information by pointing to it on the floor.

Is this teaching? It's easy to argue that it is not, but think for a moment about what the point of teaching is. When someone— anyone, not only a trained teacher in a classroom—decides to teach a learner, it's because the teacher perceives a knowledge gap and is trying to close it by offering information. This attempt suggests that the teacher has formed a ToM of the learner.

Another example of this very early stage of teaching can be seen in a study that was done with one-year-old infants but included children up to twenty-three months as well. The resulting behaviors were evaluated across three age groups: twelve to fifteen months, sixteen to nineteen months, and twenty to twenty-three months. The experimenter sat with the child in the comfort of the child's nursery school and played a fairly common game of placing a shape (triangle, circle, square) in an appropriately shaped hole. After the child became familiar with the game, the experimenter would feign an inability to place the shape in the corresponding hole and would exclaim, "Oh, I can't."

Each session was videotaped so that researchers could evaluate the child's response. The child's behavior was assigned to one of four categories: putting the shape in the correct hole, pointing to the correct hole, staring and doing nothing, or showing no interest of any kind. They found that all of the children between twelve and fifteen months put the shape in the correct hole for the adult experimenter, as did 80 percent of the children between sixteen and nineteen months. These results are similar to the previous study except that these children were not prompted by the experimenter to help close the knowledge gap (recall that in the previous example the experimenter asked the child if he or she had seen the "lost" object). In this study, children between twelve and nineteen months completed the action on their own and did not need the experimenter to offer a hint such as pointing at the correct hole. Without prompting, children performed a very basic level of teaching. They did not teach through demonstration, which would have required a higher level of sophistication, but their actions were suggesting a hint or even a general display of ToM.

The next age level showed a more sophisticated type of teaching. The children between twenty and twenty-three months pointed and uttered something in response. We can assume that they were uttering some type of direction even though their language was not yet developed. When a toddler is asked a question such as "Have you seen my pen?" and the child points to the pen and utters something, it's likely the child is responding about the pen, although he or she may not yet have the vocabulary for adults to understand what is being said.

In both of these studies, very young children were teaching! They knew where the object was or should have been placed, and the experimenter did not. By offering information and giving the experimenter this knowledge, they positioned themselves as teachers. Simple as it may seem, it's actually quite sophisticated. The little ones made a decision to close the knowledge gap between themselves and the experimenter. But *why* did they do this? The children were offered no reward. They did not know the experimenter. And children of this age are certainly not professional teachers.

A large body of research exists suggesting that children this age did so simply because they are human, implying that *teaching is an innate and natural human cognitive ability*.[24] If we can all teach, even very young children, the distinction between one person's teaching ability and that of another—that of a grad student compared to that of a preschooler, or a professional classroom teacher's compared to a bank teller's—becomes muddled. When and why, then, should we treat the classroom teacher like a professional on any level? Armed only with this understanding, it might even seem feasible that a few weeks in a fast-track program—perhaps one that trains young people to modify the behaviors of their soon-to-be students—could transform anyone into a teacher. Yet when parents think about the experience they want or expect for their child in school, few if any would settle for their child being taught by someone whose teaching skills consist of pointing to the necessary information in the manner of a helpful two-year-old. Without a doubt, a computer could fill that role more reliably and at a lower cost than any human.

But when very young children demonstrate an awareness that other people have thoughts different from their own, they are expressing a necessary and fundamental cognitive skill, one that only humans seem to have. This skill changes and becomes slightly more sophisticated as children develop, demonstrated by differences in how each age group teaches. All around us, every day, we see that humans can and do teach at almost any age. Still, what a two-year-old can teach and what a fifty-year-old can teach are significantly different, though not simply because inevitably they have different content knowledge.

Our innate ability to teach develops over time, just as our learning skills develop over time. Whereas infants exhibit a basic, concrete level of teaching (what Strauss refers to as proto-teaching), older toddlers do something more complex. One study, which was subsequently replicated, found a developmental trajectory reflecting different levels of teaching by children ranging from three to five years of age. The study began by teaching children how to play a board game. Once they learned how to play the game, they were then asked to teach a friend. The purpose of the game was to collect three flowers of different colors. There were four different colors of flowers available. Children would roll a die and pick up from the center of the board a flower that corresponded with the color of the flower appearing on the die. One side of the die had a smiley face, which allowed a player to pick up any flower, and one side had a frown, which meant that player lost her turn. They found that three-year-old children would use demonstration to teach their friend how to play the game. Interestingly, these children had low performance on ToM assessments, which by itself would suggest that they would not be able to teach because they didn't have a strong understanding of their learner. However, they were clearly able to teach. Some children even explained their steps as they engaged in the demonstration. When peers made an error while playing the game, the children would either ignore the mistake or correct it by performing the task correctly for their peers. All of these behaviors are more cognitively taxing than merely pointing or correcting, the type of teaching happening in the younger age groups.

The teaching strategies of children continued to advance in the older age groups. At the age of five, there was some demonstration taking place, but primarily the children explained the game. They would often repeat the rules when their friend committed an error: "You have to throw the die and if it shows red, you pick the red flower," "Look, the die shows red, so I pick a red flower, not blue, see?" These remarks point to a ToM—the children were teachers monitoring the progress of their learner. It turns out that children at this age also showed stronger performance on ToM assessments, which include activities that are not relegated to the artificial setting of research environments.

One could argue that when a student learns via demonstration only, the resulting responses might be pure imitation rather than the result of teaching. Whatever the case may be in that regard, the key lesson from these experiments is the undeniable difference between the patterns common to the three-year-olds' behaviors and those common to the five-year-olds. And when there was little demonstration happening and more explanation of rules and procedures, it is more likely that the learning is a result of teaching. The five-year-old teachers monitored their learners and adapted their teaching (via explanation) to the learners' progress.

As children reach age seven, they adapt even more to the learner's progress by offering more or less demonstration and explanation based on how well the learner was playing the game.[25] This more interactive, response-oriented type of teaching (referred to as contingent teaching) continues in children ages nine through eleven. Their teaching is even more advanced in that they offer the learner alternative ways to learn to play the game.[26]

Let's pause here and consider what this tells us. As we age, our natural ability to teach moves from proto-teaching to demonstration, then explanation, and eventually contingent teaching (see Appendix C). On a basic level, this means that teaching, like learning, becomes more complex as we develop over time. This is hugely important because typically we don't think about teaching as a skill (or set of skills) that changes over time, but rather something more static. We also tend not to think of teaching as something that children do regularly. Nor, even though common sense signals it to be true for people of all ages, do we often think of teaching as something that exists outside of a classroom.

Despite the importance of ToM, it still does not offer a complete understanding of what it means to teach. What it does offer, however, is one important takeaway: teaching is a human-specific activity because it requires the highly cognitive skill of ToM, which only humans have. This is great progress: humans are not just like animals!

But ToM is not the only thing that distinguishes human teachers. It is only one skill in a very complex pattern of cognitive and affective skills that operate at multiple levels. For example, simply mapping ToM to teaching does not completely capture the context-dependent interaction involved in teaching, nor the teacher's awareness of that context. A teacher using ToM might create her own image of how a child perceives the concept of slavery, but ToM would not give the teacher insight into how this perception interacts with the child's larger environment, emotions, and memory. It also would completely ignore the teacher's own contexts—her background knowledge and experience, personality, attitudes, beliefs about teaching, and ability to manage stress.

TEACHING IS A NATURAL HUMAN ACT

This ToM research presents two critical lessons that form the foundation of a new definition of teaching:

1. **Teaching is a process.** It is an interaction that occurs between humans who express a desire to connect with each other and join their knowledge. Both people benefit from the collective knowledge and the interchange, and this kind of interaction can and does happen everywhere: within and outside classrooms, with and without formally trained teachers.

2. **Teaching is a natural human act.** Teaching is a uniquely human endeavor that we employ when we want to join together and become of one mind.

Children provide the greatest example of our inherent ability to teach, and they also give us insight into our fundamental drive to do so. My nephew Jordan is four years old. One summer day, after being dragged to his older brother's football game, he decided to play a game of naming the clouds. "Titi," he said to me, "look at the really big cloud over there. What do you think it looks like?" I made the mistake of thinking this was an open-ended question, but Jordan already had a lesson in mind. After informing me it was *not* a dinosaur, he explained, "You see it's floating. Fish float. It's a big fish that takes up the whole sky, and those are the fins. Do you see them?" Jordan continued to teach me the rules of the game: he got to choose the cloud animal in the sky, and my job was to say what that animal was doing up in the sky. (An aside for curious readers: the big fish was looking for a camp where he could fit in the door, because Jordan noted that the doors in regular camp couldn't possibly fit big fish. Our newly designed sky camp had no walls, just a roof to keep out the rain. A brilliant idea, until I learned that a giraffe might want to join the big fish camp!) Jordan was teaching me because he wanted to play a cloud-naming game and I didn't know how to play. It was not a selfless act on his part.[27]

I was able to catch on rather quickly to how Jordan was teaching me to play. However, on at least one occasion, Thomas, his older brother, announced that he thought this game was "stupid." He had many questions and comments: "Why is it a fish? Fish aren't in the sky, they're in the water. Are you saying that the sky is the water? I think it's a baseball. Why can't it be a baseball?" In order to ensure that Thomas would know how to play the game, Jordan would have had to adapt his instruction. And while at age four Jordan could demonstrate, at age seven Thomas required more explanation, guidance, and reciprocity.*

Jordan was not yet developmentally able to adjust his teaching based on Thomas's needs. The result? Both children grew highly frustrated—Thomas felt Jordan was being inflexible, and, sounding wholly exasperated, he complained, "Jordan isn't letting me

---

*These developmental levels are not hard and fast, like the Piaget ladder approach. Rather, I am suggesting a developmental scale displaying how learning progresses over time. A more detailed look at this scale appears in Chapter 3 and Appendix C.

choose! Why can't I help choose?"—and decided to play with other children.

Unlike learning, teaching cannot happen independently. It must involve a teacher and a learner. As teachers, we go as far as we are able in order to close the gap between ourselves and our learner. But if we've exhausted those options and still have not reached the optimal learning dynamic, both sides tend to get frustrated and move on, mentally, physically, or both. I knew that I would receive a demonstration and some explanation of the game from Jordan, but he surely wasn't going to adapt to my learning preferences—I could take it or leave it.

But then I hijacked Jordan's teaching moment and made it my own. I thought it would be interesting to challenge Jordan to figure out how all of his cloud animals could go to the same camp that he had created in the sky. In the end we had a camp with three new design elements. One accounted for the big fish that needed doorway-free camps, the second for giraffes who needed roof-free camps, and the third for Mrs. Hippopotamus, who couldn't understand why there were no girls in this huge open camp. Parents do things like this all the time. They find small opportunities to turn an otherwise ordinary moment into a teachable moment, even if that means hijacking a lesson that began with the child as the teacher.

I first began to pick up on this early in my teaching career. In New York City many of the middle schools are located directly above elementary schools, in the same building. Space is rather tight, so you'll often find a four-story building with the first two floors for elementary school children and the top two for the middle schoolers. The younger kids get there first and usually arrive around the same time as the middle school teachers are coming in to prep. One day I watched as a father wrestled with the seat belt of the booster seat as his daughter, Eva (who years later would become a student in my class), whined, "I can't close my coat." In his rush he replied the way most of us would, assuring Eva that she could indeed close her coat. But Eva replied, "No, I can't—this is not my coat. Look, this one has holes in it!" She proceeded to explain that *her* coat closed when she pushed the side with a "nose" into the side with the hole, and that when it was done right, it "sneezed." "It's just a little sneeze, Dad. Try it—you'll see that it doesn't work." Here was a little girl teaching her dad how to close a snap on a coat. Eva described both sides of the snaps, the side with a "nose" and the side with a hole, and tried to demonstrate how they work. But her coat that day had buttons, not snaps, and she hadn't yet learned how to fasten buttons.

These instances, often referred to by teachers as "teachable moments," are the times when learning is most achievable, when the time is right for knowledge to be shared and built.[28] These are the moments where a learner provides intentional feedback on her learning (including a lack thereof or misunderstandings), presenting the opportunity for intentional teaching to follow. While all of us can, acting on instinct, seize those naturally occurring moments and give student-centered responses, adults often have the skill to actually *create* intentional teaching situations. This is the level of teaching that Jordan was missing when he was teaching me and his brother the cloud game.

In Eva's case, the father processed the feedback he was getting from Eva and responded in a way that took into account her individual needs. In essence, he formed a theory of her learning brain and then considered how he should respond in order to intentionally teach her something that she needed to learn. Rather than just buttoning Eva's coat for her, her dad leaned over and responded, "That's because this is a button. It doesn't have a 'nose' to snap in, so it can't sneeze. It must be frustrating not to know how to use those holes on the other side. Would you like to learn how to button your coat?" Eva's dad began his teaching moment by doing three important things:

1. **Respecting existing knowledge.** Eva's father acknowledges that Eva does know how to close her coat with snaps.

2. **Responding to a need for new knowledge.** Eva is frustrated that she does not know how to close a coat with buttons.

3. **Offering agency.** Eva's dad provides her with the autonomy to choose whether she would like to learn this new skill.

Eva's father did not take on the role of a transmitter of knowledge (as in the empty vessel approach). Rather, he saw that Eva already had knowledge, needed new knowledge that built on the knowledge she already possessed, and had the ability to decide whether she wanted to receive new knowledge.

If Eva was too frustrated at the moment to engage in this teaching-learning interaction, her dad might have motivated her by asking, "Why might it be a really good idea to close our coats before we go outside?" in an attempt to interest Eva in learning how to button her coat. To continue promoting Eva's cognitive development, he might have begun to brainstorm with her about other things that require buttoning, or he could have asked Eva how buttons are different from snaps. The capacity to find these prompts and motivators comes from an understanding of Eva's interests, her personality, and what methods are most successful for getting her to learn something she finds frustrating. However, it's crucial to keep in mind that motivating someone to perform a task you want him or her to do is not the same as motivating someone to learn.

PET scans and fMRI scans have the ability to reveal that learning is occurring by showing blood flow and oxygenation patterns that indicate activation of certain parts of the brain.

But we don't yet fully understand the actual processes of learning that take place in the brain. At this point we can only see phenomena that we believe correlate with learning taking place in the brain.

Fortunately, this was a good day for Eva, and she showed interest in learning how to button her coat. Her dad then responded, "I wonder how we could get this round button to fit into that slit on the other side of your coat. Why don't you tell me what you tried and why it didn't work?" Her father's method of contingent teaching addressed Eva's need to feel like she was being heard. Like many first graders, Eva wanted to feel empowered to display her knowledge, so this method was just right for her.

These types of interactions happen all the time, quite naturally. Each teachable moment looks different from every other, but children being taught by their parents, or other similarly situated adults, is one of the most common teaching relationships. Looking at why parents teach is yet another path that leads to explaining why anyone teaches. On a very basic level, we teach so we can belong, and so we can advance together as a group. A parent helps teach a child how to walk not because we must walk upright to survive but because it is a social norm, something society values. We teach reading and writing not because humans are unable to communicate without them but because they allow us to become part of each other's context.

Past research has leveraged our rich understanding of learning to frame our exploration of teaching and a basic understanding of its cognitive processes.[29] However, in order to truly understand the interaction between a student and a teacher, we must focus just as much effort on understanding the teaching brain, which in part is responding to the learning brain. We've come a long way indeed, but the need for more research that explores the ontogenetic and phylogenetic origins of teaching—from its early development in toddlers to its refinement in the expert, veteran classroom teacher—is still great.* By acknowledging teaching as a natural human cognitive skill that demands interaction, research and policy efforts have the opportunity to shift their focus toward truly understanding teaching across brains of all types and ages in an attempt to better comprehend the complex social system in which we educate one another.[30]

---

*Ontogenetic refers to the development of a particular individual from her first immature stage to an adult. Phylogenetic describes the development of a group of related living things by evolution over a long period of time.

2

COOKIE-CUTTER SOLUTIONS AND OTHER MISSTEPS OF EDUCATION REFORM

Close your eyes and envision a teacher teaching. What does it look like? In all likelihood a teacher is standing at the front of the room, facing rows of desks; students are raising their hands, and information is being transmitted and received.

We could say that this imagery is what we typically see in movies or commercials, but I'd still ask why this is *your* vision of teaching. Are humans born awaiting the artificial setting of a classroom to begin teaching?

Humans start to teach almost immediately, without ever having gone to school. Yet how we teach in most schools is far from natural. It comes from definitions built upon research, much of which, as discussed in chapter 1, is grounded in animal studies. Much research and education reform continue to be designed based on that misguided, incomplete foundation. At its core, our institutionalized manner of educating does not match how humans teach.

Let's look at some of the most common—and controversial—examples of education reforms. The No Child Left Behind Act (NCLB), passed in 2001, was a critical piece of legislation that shifted how education was practiced in this country. The belief driving this legislation was that schools needed to demonstrate that they were helping children progress and achieve higher levels of learning. In order to do this, they tested students more frequently and rigorously. State exams became mandatory and were considered evidence of the achievement of the student, teacher, the school, and the district overall. The focus was on math and English language arts (ELA), with the goal being for all students in public schools to reach 100 percent proficiency levels as measured by annual standardized tests.

While one of the purposes of NCLB was to close the achievement gaps between low- and high-performing schools, the pressure was equally on teachers, who were suddenly expected to adjust their teaching in order to prepare their students for these tests. In 2001 I had finished my fourth year as a teacher in the New York City public schools; I had just moved to a new school and was again going to be teaching humanities, English, and social studies combined. But according to the new state mandates aligned to NCLB, only one of those subjects mattered. As a middle school teacher, I was also teaching the two grades that mattered most: seventh and eighth. Students wishing to attend a particular high school would send their seventh-grade test scores to prospective high schools in a process that functions similarly to the matching that medical students participate in as they search for a residency placement. Students choose their top five schools, and schools identify their top student choices. On match day computer programs compare these lists and pair up students and high schools. The pressure on seventh graders and their teachers was high. And it didn't let up after students were matched. Eighth-grade test scores were used by the state to judge each school's overall performance, so the administration—in each school and across each district—focused much of their energy on increasing eighth grade scores.

I remember vividly the day my principal informed me that I was required to stop my regular course of teaching for two weeks in order to train my students for the ELA exam. This had to be done for both seventh and eighth grades, because that year policy makers had decided that schools would also be judged on the extent to which seventh graders were matched to high-performing high schools. The middle school administration proceeded to alter the entire school's schedule so that students would receive two weeks of full-day test preparation, complete with a shortened thirty-minute lunch and no recess.

According to my principal and NCLB mandates, my job was to *train* these students and, in turn, increase their test scores. Teachers were being "held accountable," and student scores would prove teachers' effectiveness in the classroom. It sounded something like, "Vanessa, you must ensure your students have all the necessary information to score well on the exam." I felt extremely proud as the giver of information, the keeper of sacred and powerful knowledge. My students needed me. Like a super-hero, I could save the day!

There were two ELA teachers assigned to each grade, and the students were divided into two groups. As a four-year veteran, I was assigned the students presumed to have the greatest potential to score well. The exam is scored on a scale of 1 to 4, with 4 being the highest. I was given mostly students expected to score at level 3 or 4, along with some mid-level 2s. The administration took the stance that low-level 2s were not worthy of extra support because they were not likely to be able to move their score up enough to benefit the school. (Dissecting the evils of this reasoning would require a lengthy tangent; suffice it to say that this mode of thinking is terribly harmful and dehumanizing to both teachers and students.)

Each day for two weeks I was in a room with approximately forty students, as opposed to my usual twenty-eight. The assistant principal came in to ensure I was working through everything that students needed to know in order to do well on the state exam. Every hour was mapped out strategically. I explained carefully to students how to read the directions, because clues to the correct answer were hidden in the directions. We reviewed countless reading passages to identify the genres to which they belonged. The students mastered exactly how long different types of questions should take to read and respond to, as well as how to easily garner the maximum amount of points. Scattered in there, of course, was standard middle school English content (i.e., genre, plot, sentence structure, writing and reading skills).

I called this test prep training, but my administration and many others insisted that it was teaching. In fact, what I was doing was very much in line with a behaviorist definition of teaching, one that directly supports the efforts of a mandate such as NCLB: train the students so they perform better on tests, and measure teachers' accountability by the performance of their students. I altered my natural teaching behavior in order to promote a particular type of learning—how to take the state exams—in my students. I was the transmitter and my students were the receptacles.

The harm to students in this scenario is evident. What is less apparent is that this dynamic offered no benefit to me, the teacher, either professionally or personally. There were no relevant or connected entry points around which to design the lesson or which I could use to underscore my strengths and interests, including my understanding of the learners. Instead, each day I felt like a drill sergeant. With six crates of hanging file folders full of detailed lessons, teaching scripts, and practice sheets and tests, in no way was I teaching based on the individual learning needs of my students. I was given clear instructions based on desired outcomes defined by officials on high. There was never any cause for discussion. There was always only one right answer and I was the keeper of it. Who I was didn't matter in this interaction, nor did it matter who my students were. It was pure skill and drill. I could have been any teacher and the students could have been any students. The lessons would have been done in exactly the same way.

Did the students learn? Well, by the end of those two weeks they were scoring better on the practice exams than they had when we first began their training. Their scores on the actual test also increased from the year before. But I suspect even Pavlov's dogs would have scored better with all that drilling! In essence I was using the same techniques as a mother pied babbler. Neither I nor the students were huge fans of this type of teaching, but then again neither are we birds or dogs.

Modern education reforms tend to favor what amounts to animal teaching over forms of teaching that are more sophisticated and specific to humans. Teachers are given report cards primarily based on their students' standardized state exam scores. Teaching built on behaviorist principles is incentivized in order to reach certain objective goals, what we call standards. If a teacher can more efficiently increase student test scores through basic test training, it's no wonder that many would choose not to bother with more complex forms of teaching. After all, aren't the teachers doing it for the children?

However, as discussed in the previous chapter, the concept of teaching as primarily a selfless act meant only to benefit the learner is fundamentally flawed by its unidirectional character. Devaluing or ignoring the teacher's context as part of the process does not acknowledge the act as a dynamic interaction. It's high time to ask ourselves if we are investing our time and money in education reforms that offer actual evidence of success in humans, as opposed to animals.

We've looked at various responses to the question of what teaching is, and have begun to unpack what's wrong with the usual responses. To judge the quality and validity of large-scale policies and reforms, we must ask: What basic assumptions do we build from? Do we design and evaluate teaching based on both our knowledge of teaching and our knowledge of learning? Is teacher evaluation based on our ability as humans or the less sophisticated ability of animals? Specifically, is it based on assessments designed to actually measure the teaching of humans? And finally, what is the nature of the connection between learning and teaching?

FROM THE MIXED-UP FILES OF CHARTER SCHOOLS

Let's consider another example of how modern teaching reform in the United States has been built upon behaviorist models of learning and teaching. A large number of charter schools have sprung up in cities across the country to offer parents alternatives to the mess of public schools and their reliance on standardized tests to measure learning. In general, this is not an unreasonable response. However, all charter schools are not created equal. The approaches to teaching embraced by many large networks or chains of charter schools provide serious cause for concern, or at the very least cause for inquiry.

Today there are several teacher training programs targeted at charter school placement. Programs such as Teach for America (TFA) typically recruit high-achieving college graduates with an eye to filling the teaching gap in low-income schools. Over an average time span of just five weeks during the summer before they begin teaching, these ambitious young people essentially get boot camp training in state education mandates, disciplinary methods, and curriculum planning. The training regimens are generally extremely rigorous and rigid. Once they begin teaching, the novice teachers receive continuous feedback from mentors and principals who are often former graduates of the same program. On average mentors and principals of these new teachers have four to five years of experience. In general, organizations such as TFA draw upon a market of young people with no previous teaching experience or background in education. The charter school systems offer to train these young people in exchange for at least a two-year commitment to the schools where they are placed. Many teachers who begin in a program such as TFA stay at their schools anywhere from two to five years.

The trend may seem new and confusing to some, especially those who lived through the era in which baby boomers were taught by lifelong teachers. However, this is not a new concept. Before teaching became primarily a female-dominated profession, young males would go into the field for one or two years to gain practical experience before moving on to a more prestigious profession such as medicine or law. The belief was that these young men could easily teach children because they were intelligent and motivated—a rationale that's remarkably similar to what drives the models that have been expanding over the last decade or so.[1]

Following their condensed, rigorous, and inevitably limited training, how do graduates of programs such as TFA teach? A *New York Times* article published in August 2013 illuminates this quite well. The article describes how, after observing a couple of first-year math teachers reviewing place values with their middle schoolers, the young principal interrupted their instruction to point out that one of the teachers had incorrectly instructed her students on how to engage in the school's celebratory chant. The principal exclaimed, "It's two claps and then a sizzle." The chain of charter schools in which the principal and teachers work has strict guidelines for how students should dress, walk in hallways, enter classrooms, and even celebrate. Both teachers and students are trained on how to correctly implement all actions. In another example given in the article, a teacher mentor with four years of math teaching experience was in his biweekly forty-five-minute mentoring session with a novice art teacher. After commending her for giving students helpful drawing tips, he advised her that students needed to sit down at their desks more quickly after entering the classroom. The novice teacher was asked to role-play four times how she would begin class with the new time-keeping app that she was instructed to use in order to give students precise deadlines.[2]

Models like this are directly in line with the same animal-behavior-rooted ideas about teaching formed in the 1960s—which should prompt us to wonder why this approach is considered innovative, and furthermore why it is such a popular and well-supported effort in education reform. I would suggest that it's become popular because it's a clean, simple, precise input-output model offering direct training guidelines and quantifiable outcomes. These features make it not only easy to "teach" teachers but also much easier to explain to frustrated parents and citizens—all those folks with jobs, hopes, few tax dollars to spare, and understandably little patience for digesting the nuances and complexities of teaching and learning. But as sexy and appealing as simple fixes and sharp sound bites can be, if we are really going to construct an "accountable" education system we can't afford to be so narrow and ill-informed in our thinking.

We know that when we use this method we can put teachers through fast-track training programs and they will come out with information on teaching (or training, depending on which definition is in play). They will use that information to transmit specified sets of knowledge to students, who ostensibly will report it back on exams. These exams can then be used to measure students and teachers alike. But these exams are designed to measure narrow areas of students' content knowledge and skill development. They were *not* created to directly measure teaching ability (behaviorist or otherwise), teaching effectiveness, or teachers' skill level. Utilizing student test scores to measure teaching is like using only a thermometer to determine whether you have the flu. A thermometer may tell you something about your condition, but it isn't designed to identify viruses, assess congestion, or evaluate levels of lethargy. We turn to other measures and the expertise of doctors for that, because that's what it takes to get a complete and accurate picture.

Training high-achieving students to become teachers is not in itself a bad idea. And indeed, charter schools are not all bad.

They do offer some choice in sometimes dire situations, and they also offer hope—something we can hold on to, examples of how we might fix what we think is broken in our education system. Charter schools also give us a seemingly straightforward way to close the achievement gap by focusing predominantly on boosting test scores. By training students to do better on standardized exams, many charter schools are then able to claim that the achievement gap has been diminished. But as we know, outcomes on standardized tests don't necessarily reflect deep learning. Charters may also offer us a salve for the collective guilty conscience associated with the reasons for that achievement gap existing in the first place.

We've accepted that teachers are the most important component of student success, so supporting them with these techniques gives us a way to guarantee good teaching . . . right? And good teaching leads to good learning, good learning to good jobs, and good jobs to dominating the world—oh, and closing racial, achievement, and economic gaps. It's a foolproof plan, right?

Not so fast. If teaching is simply a matter of training and following a basic input-output model of learning, that leaves us equipped to approximate teaching up to the developmental level of a chimp, dog, or bird. This type of teaching does not come near the potential complexity of human cognitive skill.

Let's again consider the fact that the rudimentary teaching skills of the input-output model develop early. In the toddler years humans have teaching skills that are very similar to those of animals, like pointing and demonstration. We offer simple explanations of how things work: you can get to the top of the slide by walking up the ladder. This kind of modeling is a form of teaching, but it's a *simple* form of teaching—actions that are utilized in input-output models of teaching. However, as humans develop over time, we evolve past this animal-like style of interacting and our teaching becomes more complex. Just as we develop emotionally, socially, and intellectually over our life span, so too does our teaching ability; it grows and adapts each time we are placed in a new context.

We do not expect that the teacher in front of a group of three-year-olds is presenting content with the same level of complexity as that seen in a university professor teaching a seminar of graduate students. However, *interacting* with the preschoolers—recognizing those learners' level and responding appropriately—requires just as much cognitive thought as it does for a professor to interact with adult graduate students. Some may argue, in fact, that it is easier to interact with learners at a similar cognitive level as oneself, such as in the graduate context. It is indeed quite difficult to teach learners who are at a much less developed cognitive level.

A rookie teacher in a formal or classroom context is in a parallel situation to a toddler in the nascent phase of learning to teach. Both may have been able to master demonstration and explanation—the teacher through rigorous training, the child as a natural consequence of growth and development—but teaching that is actually based upon how individual learners respond develops only through experience. This skill is called contingent teaching: when the teacher knows how to proceed in teaching based on the learner's response. If the learner has gained the knowledge taught, then the teacher can offer less support; if the learner has not yet mastered that knowledge, the teacher can provide more support.[3] I estimate that a novice teacher can come to a basic understanding of contingent teaching within the first year of meeting her students. Just as a learner moves back and forth between a functional level of understanding and an optimal level of understanding, so too do teachers: teachers develop their skills not all at once but systematically, through experience.

Strauss and Ziv refer to the evolution of contingent teaching as the creation of "mental models" of the "dynamic workings of pupils' minds when learning occurs."[4] In other words, contingent teaching is based on the teacher's ability to develop a theory of mind (ToM) about the learner. The researchers go on to explain that there are two types of mental models:

1. **Espoused.** Espoused models are those that can be constructed based on the ways people speak about their teaching.

2. **In-action.** In-action models must be inferred from people's actual teaching.

In-action mental models have been revealed through studies in which researchers videotaped and analyzed what was happening as teachers were teaching.[5] Their findings show that teachers' in-action mental models comprise the following:

1. **Cognitive goals** that teachers want their students to achieve

2. **Cognitive processes** that teachers think lead to these cognitive goals

3. **Assumptions** about how teaching in a particular way leads to these cognitive processes and, in turn, the cognitive goals

4. **Meta-assumptions** about learning and teaching

This approach treats the complexity of teaching not only as a way of interacting with learners but also as a process unto itself. It thus makes it easier to see how challenging it is for novice teachers to develop in-action mental models of their students, and moreover how unlikely it is that adequate training to prepare for this process could take place in a five-week summer program. It is also unlikely that even after three years of teaching, a teacher will have mastered how to practice this complex cognitive task within the context of a classroom. As we will see, these skills are indeed what teachers aspire to build and develop over the course of their careers.

WHAT ABOUT "BRAIN-BASED" PROGRAMS?

Still other education reforms tout "brain-based education" strategies that draw upon the very new—and rapidly changing—field of neuroscience. Though they take the individual needs of learners very seriously, many "brain-based" programs give the impression that we actually know a lot more than we really do about what is going on in the brain. Despite the amazing research that has been done into how the brain works, we are only just beginning to interpret the patterns that emerge from this research and link them to the vast variability that we know exists among individuals in the process of learning.

Don't get me wrong. I have sincere respect for work in brain-based education, as you might have guessed from the title of this book. But there are large and real dangers in oversimplifying the research from neuroscience and applying it to education so concretely. What we do know is quite extraordinary and revolutionary, but it is yet incomplete. We know that understanding the mechanisms of the brain is important. We know what areas of the brain "light up" when we accomplish certain tasks. Most significantly, we know that each person's brain is dynamic, variable, and dependent on one's context.

The premature push for certain brain-based practices to be implemented is hardly surprising. It is actually fairly typical for researchers of all sorts to make suggestions for how to improve education, regardless of whether their research has directly involved schools, students, or teachers. In this case, a number of researchers have called for a more brain-based approach to the classroom, concluding that because our brains are all individually wired, each student should have individualized instruction.

One very clear example of how brain research leads to suggestions about how to improve education is offered by John Medina, a molecular biologist. He says that because studies have concluded that the brain cannot multitask (it only holds attention for about ten minutes and recognizes patterns better than details), teachers should teach in ten-minute segments, beginning by explaining the entire topic in one minute and connecting it to a core concept. Each ten-minute interval should then end with a hook that gives the learners enough emotional arousal to interest them for the next ten minutes. Additionally, because repetition is helpful to our long-term memory, schools should cycle students through the same lesson three times in one day.

At first glance Medina's logic and his suggestions for translating brain research into teaching practice might seem quite rea-

Studies have concluded . . .	Therefore in schools teachers should . . .
We cannot multitask	Focus on only one task and topic at a time
We cannot keep our attention focused for more than ten minutes	Teach each topic and task in ten-minute intervals
We can recognize patterns better than details	Explain the entire topic in one minute and connect it to a core concept
It is better to map information to our brain when we have increased attention and motivation	End each ten-minute interval with an interesting hook
Repetition is helpful to long-term memory	Cycle through the same lesson three times per day

Table 1

sonable. But it's less than sensible, if not outright alarming, for experts to call for research to be applied to practice without taking the time to think through what such an application might require. This is why research studies often seem far removed from our everyday life.

It's worth noting that the brain studies on which these recommendations are based were not performed on teachers, nor do they examine teachers' interactions with learners. (Nor, of course, were the foundational studies on teaching outlined previously, which were based on animal research.)

Painting a clearer picture of what the list above would look like in practice is instructive. The average K-12 teacher's day consists of teaching five classes that run forty-five to fifty minutes each. This brain-based approach would require five distinct topics to be taught in one fifty-minute period. Each topic would have its own one-minute introduction and culminating hook. But there is no room in the standard day for the lesson to be repeated three times. Perhaps schools could divide their classes into twenty-five-minute periods and a student would have three English class periods, three math class periods, three history class periods, and so on. Each class in a subject would consist of the same exact lesson taught in the same exact way. Students would arrive in the class and get settled, the teacher would deal with school announcements and class follow-up, the lesson would be taught (with students perhaps scrambling to engage in various groupings for peer collaboration, because research has shown that to be beneficial too), and students would have a chance to practice so the teacher could evaluate their progress (perhaps squeezing in some claps and then a sizzle). This would all take place three times each day. Imagine this, for both the student and the teacher, as an actual school day.

In spite of the novelty of brain-based initiatives in the field of education, their basic foundation still reflects a traditional behaviorist viewpoint boasting a linear input-output model. Even in this cutting-edge brain-science-based model, the teacher remains the giver of information, holding knowledge that is transmitted to and received by the student—although in this example it happens three times. The teacher's behavior must be altered specifically to impart this information, for otherwise who would teach the same thing three times in exactly the same way in one day? Only if we think that teaching is a selfless act in which teachers are merely tools for learning would anyone propose such an abnormal routine. Ultimately the design involves just demonstration and explanation, which makes for effective training but not necessarily deep learning.

Is this the best we can do? Most research in the learning sciences, including brain research, offers some suggestions for education, but they have all focused on the learner. That's not a bad thing, but it's only half the equation. Teaching always involves a learner—otherwise it cannot be called teaching. A complete, accurate, and up-to-date theory of teaching must be based not just on theories of the learner but also on a more precise understanding of the nature of teaching. Only by understanding the *connection* between learning and teaching can we understand the dynamic interaction of teaching. Perhaps by understanding the learner's brain, we can come closer to what the teaching brain is all about.

PART TWO

IT'S ALL ABOUT SYSTEMS

PART TWO

IT'S ALL ABOUT SYSTEMS

3

UNDERSTANDING THE
LEARNING BRAIN

I was a fairly experienced teacher, with almost eight years of teaching under my belt, when I encountered Adam. I had met Adam's mother before the school year even started because she was the president of the Parent-Teacher Association, and she made sure to tell me that Adam was "different." "He has a lot of energy," she had said when we first met. "He's a great kid, though, and I'll make sure he does his work."

As any experienced teacher knows, the president of the PTA is not someone you want to disappoint. However, I didn't quite know what she was telling me. Was she making excuses for some heretofore unaddressed behavioral problem? And what did "different" mean, anyway? After all, in many ways my classroom was different. Even though I taught seventh and eighth grades, my room had a rug that created a meeting space, as well as a couch. We had books in color-coded bins placed to create different types of reading libraries; pillows and stuffed animals were in baskets all around each reading area. The desks moved around the room to suit our needs and were never in rows, unless perhaps we were doing a mock press conference that warranted a more realistic setting. So maybe Adam would fit right in.

It became clear in the first couple of weeks of class, however, that for Adam, this level of flexibility was not an ideal learning environment.

"Adam, is it difficult for you to focus when we have our mini-lessons?" I asked him one day when we were in a one-on-one workshop together. (Mini-lessons are fifteen-minute lectures that often start a class session when new content or skills are first introduced.)

"I'm sorry, I'll pay more attention!" Adam quickly replied.

It was pretty clear that Adam was used to this type of feedback and had an instant reaction of defense or protection, depending how you look at it. But what I had seen in the first few weeks of school led me to believe that Adam actually *couldn't* pay more attention—at least not by exhibiting the behaviors we typically associate with the act of paying attention. It wasn't just his body language during mini-lessons that clued me in; it was how he interacted with his friends, how he shared what he was interested in outside of school, and how he worked with his peers during class activities. Everything about Adam suggested that he was bright, motivated, and successful—just not typical. So it turned out Mom knew best: Adam *was* different.

As we talked more, Adam began to share more comfortably. He revealed that he just needed to move around. "Sometimes I just can't sit still. It's not that I'm not paying attention; I just can't pay attention if I'm not moving around. It helps me focus."

This wasn't the first time I'd heard something like this. As a student teacher, I'd had a student who said that she needed to scribble on her notebook in order to process what was being said. She didn't scribble anything in particular; she just put pen to paper and constantly kept the pen moving. Not knowing any better, I had made the mistake of forcing her to stop many times until I finally gave up. That experience had left me feeling insensitive, undifferentiated, and like someone who didn't even stand by her own rules (no scribbling on the journals). At that point, I had still been a novice student teacher without a strong sense of who my students were as individual learners. I'd also understood very little about how my own context affected my teaching.

But now with Adam, I had a strong awareness of who my learners were as individuals. I had a sense of Adam's cognitive abilities and emotional state, including his theory of mind. I was also aware of how I personally felt about him and my teaching. In addition, I had a strong sense of the school context—a small progressive school that allowed necessary adaptations to improve student-teacher interactions—and I knew it would help us connect with each other. It seemed I had all the ingredients to find a solution; the challenge before me felt totally surmountable. In short, I wasn't going to make the same mistake with Adam that I had made with the doodling student several years prior.

I opted to be direct. I asked him, "So what typically works for you when that happens at home? How do you focus?" I was not saying I could create the ideal environment that Adam had at home, but I was hoping that we could figure out a compromise. What we worked out was that when he needed to move, Adam could manipulate his Koosh ball, a type of soft, rubbery stress ball. Or he could get up and stand in a different area of the room, or go sit on the couch. Although we had rules about how this would happen, I was giving him the freedom to move when his brain and body seemed to demand that kind of action. This arrangement required trust and a real understanding of each other.

There were days when I knew Adam was pulling my leg and, like any other thirteen-year-old, was trying to get out of work. There were other days where he wanted to do a headstand off to the side of the room, and that was just fine with me. I felt confident that the classroom experience for both me and Adam was better for this agreement because a better learning environment for him is a better teaching environment for me, and vice versa.

I don't have some wonderful riding-off-into-the-sunset ending to share about Adam. The situation wasn't perfect, and I can't claim that I was always the perfect fit for Adam as a teacher.

But I can say with confidence that we worked at it together, we reevaluated, and readjusted. In some classes teachers responded by tightening the reins and demanding that Adam change. In other classes teachers let Adam do whatever he thought was best and they followed his lead, working overtime to meet his needs. I didn't do either. The truth is that I couldn't design my class teaching around Adam. I had thirty other students, each with their own learning style, many of which were completely different from Adam's. Nevertheless, Adam knew that I cared, and not just in what I call a "save the children" way, but in a way that spurred me to do my best to address his needs whenever possible and to call him out when he was trying to put one over on me. This made our relationship honest and real.

Adam didn't necessarily do any better in my class than he did in the classes of his other three teachers that year—he always figured out a way to get respectable grades. However, I do know that his experience in my class was significantly different: he didn't scream and fight with students in my class, his mother didn't send me nasty e-mails wanting to meet with me all the time, and Adam didn't disrespect me. We just talked honestly, and because of that my class created less anxiety for him around his differences as a learner. And in reality Adam made it easy, because he was the kind of learner who couldn't help but let you know when your teaching wasn't working for him.

It's customary to act as though cases like Adam's are the most difficult ones, but the truth is that they are the easiest. It's like saying a baby would be easier to take care of if they didn't cry. The only way we actually know what to do and what not to do with infants is because they *do* cry. The students who are most difficult to teach are those who don't speak up when the teacher is doing something wrong. They smile and act like everything is fine, and usually their grades suggest they are moving along reasonably well. In some cases they might be shy, and so we don't notice them much.

So how do we understand teaching from the perspective of a learner's needs?

SHIFTING THE LENS: UNDERSTANDING THE VARIATIONS IN LEARNERS

As a student, I was pretty difficult to teach. While my teachers readily observed that I talked too much in class, none of them ever seemed to notice that I was struggling. It's not that I didn't tell them, but they had a hard time believing it. Generally I got good grades, mostly A's, and in high school I was on the honors/AP track. Still, I insisted that tests were often difficult for me, that reading took me a really long time, and that although I could talk about complex themes in literature and history, I struggled to actually explain my thoughts in writing.

Since I performed well on the tests, no one saw any real problems. What was not apparent was the cost to me of doing well on all those tests. I imagine teachers may have assumed I was exaggerating or was just high-strung. For my part, learning had become torture. I can't recall a single night before an exam when I didn't become ill, often to the point of throwing up. I felt sick nearly every Sunday night just from the thought of going to school the next day. Is it a successful learning experience if an A on a paper comes only after a month of sleepless nights, or if a 100 on a test is the grand prize for puking? My suspicion is that most teachers—indeed, most people—would think not. More to the point, what do grades tell us absent any awareness of the actual experience of learning, of the process that a learner undergoes to earn those marks?

Struggling learners are not just those who are failing courses or causing a scene. Learning struggles and challenges come in all shapes and sizes. We all learn differently, and understanding those differences opens a pathway to comprehending what it is we need to know about learning. Indeed, the biggest, most important takeaway from all the recent research in the learning sciences—cognitive, pedagogical, and neuroscientific—is the incredible variability of how people learn.

LEARNING AS A DYNAMIC SYSTEM

Throughout the last thirty years, but most notably in the past fifteen years, researchers have ushered in a new and more comprehensive model for how learning actually takes place in the brain. The thrust of this new model is based on the concept that the brain (along with the body) functions like a dynamic system.

The term "dynamic system" comes from physics, and refers to the complex interplay of any phenomenon and its environment. It's in the same family as chaos theory. Looking at how storms develop offers an accessible way to grasp the concept of dynamic systems. Recall the 2004 tsunami in the Indian Ocean. The first images we saw in America were of the storm ravaging the popular tourist destinations of Thailand. At first glance it looked as if out of nowhere a massive wave engulfed the beaches of Phuket. Tourists and locals were caught off guard, and people watching on television were shocked at how such a massive natural disaster apparently could happen in a split second, with no warning. However, that was not the case. The tsunami was caused by an earthquake of magnitude 9.0 under the Indian Ocean.

For anyone hazy on their middle school science, here's a quick recap. Tectonic plates usually glide along the earth's mantle, but sometimes they bump into each other or get caught on each other. Hitting another plate or suddenly releasing the pressure can cause an earthquake. But the earthquake is the culmination of a very long process. The earthquake that occurred in 2004 was the result of plates that had been pushing up against each other, building up pressure, for thousands of years. Finally one of the plates won the battle and slid over the other, causing a massive quake. As one plate rose above the other underneath the ocean, it literally pushed up the water above it. The wave created by this push consumed the coastline of eleven countries. The tsunami itself was an intricate pattern connected to each factor that contributed to its creation. Taken as a whole, the process is known as a dynamic system: one event triggers another, which triggers another, and so on. Events occur sequentially and in response to the environment at several different levels.

Even though there were unique factors in play that made it unlikely that anyone could have specifically known when and where the 2004 tsunami would occur, the pattern was recognizable. Such patterns can be of tremendous use in helping us to predict future events. And humans like predictability. I remember that when I talked to my seventh graders about the tsunami, they were terrified—they didn't like feeling as if disasters could happen at any moment with little or no warning. They'd dealt with that three years earlier, on 9/11, and it was a feeling they didn't want to revisit.

On the first day of that school year, I had promised my students something different: "Learn your history and you can predict the future." Such an assurance may seem like a cheap trick to play on students, but it wasn't a trick. I'll admit it helped get them excited about studying history, which tends not to be a favorite among adolescents. But I made the promise because I do believe that we can predict the future when we study the past. However, predictions are just that, so I was careful not to imply that predictions are certain to come true. The point was that studying history is about recognizing patterns, knowing when those patterns have appeared before, and understanding their impact. Being able to recognize patterns and predict outcomes— like what events might result in a tsunami—helps us to feel safe, and even to develop a plan of action to move forward with some level of confidence.

Our bodies are also dynamic systems. For example, we know there is no one cause of diabetes. The condition is triggered by multiple factors, including being chronically overweight, living a sedentary lifestyle, and a genetic propensity toward poor blood sugar control. Doctors and scientists know that diabetes might be avoided if you target one or more of these factors. But it's the entire *system* of factors that causes the disease. It's all connected.

In much the same way, the brain functions like a dynamic system. The way it senses, processes, and responds is the result of its environment, both internal and external. The brain does not function in isolation; rather, it is constantly changing and responding to the various and unique contexts in which we place our bodies.

LEARNING IN THE BRAIN

Learning occurs because our brain is in constant interaction with its environment. Each student brings biological factors into a classroom; each student has natural cognitive strengths and weaknesses, including working memory (taking in information, manipulating it in one's mind, and understanding its significance), declarative memory (facts and knowledge that can be consciously recalled), the multiple facets that make up executive functioning (manipulating information in one's head, shifting from one task to another, and relying on coping skills to manage emotions and stress), and overall ability (plus the ability on any given day) to manage emotional responses.

In addition to all of these factors, the physical environment of the classroom, the curriculum, and the student's home life are all contextual influences on a student's learning process and performance. Each learner has a unique mix of these internal factors and applies them in a unique pattern when he or she is learning something.[1]

Understanding the variability of learners is not as simple as distinguishing among different learning styles (visual, spatial, auditory, verbal). Nor is it a loose translation of Gardner's multiple intelligences theory (that people show varying combinations of strengths and weaknesses in areas such as logical thinking, spatial reasoning, kinesthetic awareness, emotional intelligence, and so on). Rather, the variability that shows up in brain research is based upon the fact that learning is a dynamic, interactive, context-dependent process. An individual learns distinctly because of her own individual biology (including genetic inheritances and overall physical health), access to good nutrition, the learning environment, engagement with different kinds of learning tools (books, websites, diagrams—the list is endless), and teachers. Learners are in constant interaction with other external influences, such as their friends, family, culture, and society. Together, these internal and external characteristics interact to form the learning brain.

This new model of learning boils down to four key concepts:

1. Learning is dynamic and changes over time.

2. Learning is both cognitive and emotional.

3. Learning is context dependent: some children will learn better in certain situations, with certain supports, and learn less well in others.

4. Learning is interactive: it is a social enterprise that happens in concert with a variety of active factors in the learner's environment, including teachers, parents, peers, textbooks, apps, and so on.

This last concept reinforces why the empty-vessel theory is so inaccurate. If learning were a one-way street, with a teacher depositing information and knowledge into a student's brain, the student would not actively be working or responding to that knowledge. She would simply be taking it all in and trying to remember it, as if learning were that simple (or boring).

DYNAMIC SKILL THEORY

Kurt Fischer extended the work of Esther Thelen's dynamic systems theory and was one of the first researchers in the learning sciences to apply the concept of dynamic systems to learning.[2] Fischer developed a unique way to understand how learning happens as a developmental process, which evolves over time from simple and basic to more complex and sophisticated. Fischer calls this learning framework dynamic skill theory, and he based it upon research that shows a ten-level skill scale, illustrating how a learner becomes more competent and develops more complex understandings over time.[3]

As Schwartz and Fischer point out:

> Dynamic skill theory offers an adaptable framework for analyzing the learning process in various contexts. The ten levels that comprise the last three tiers are well-suited for characterizing and measuring changes in cognitive (and emotional) learning observed in students. (Reflexes enter the learning picture only rarely; most learning after early infancy involves actions, representations, and abstractions.) Within each tier, people develop through four levels of increasing complexity and thus attain mastery of skills at that tier. The first level is a single expression of the understanding that is characteristic of the tier—reflexes, actions, representations, or abstractions. The second level is a coordination of single expressions into mappings. At the third level, individuals coordinate two mappings to form a new and more complex skill level—systems. The fourth level of a skill demonstrates the ability to coordinate multiple systems into a system of systems, and this achievement forms a new structure to begin the next tier: Reflexes form actions, actions form representations, and representations form abstractions.[4]

Dynamic Skill Theory

Ages indicate approximate time at which that cognitive skill level emerges. Tiers are listed on the left, and levels within each tier are listed to the right of them.

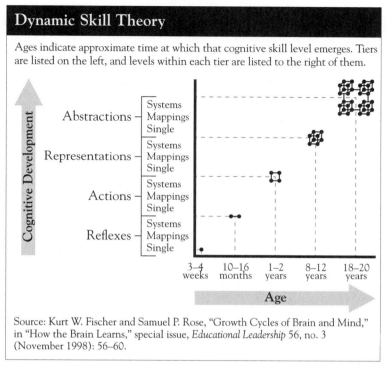

Source: Kurt W. Fischer and Samuel P. Rose, "Growth Cycles of Brain and Mind," in "How the Brain Learns," special issue, *Educational Leadership* 56, no. 3 (November 1998): 56–60.

Figure 1

According to this theory, learning is a skill that we're all born with that develops over time. When we are first born, we learn from our body's natural reflexes. Our eyes, ears, arms, and such just react, and we learn from that. Of course, learning can be more complex than that. The distinction is that this earliest type of learning is not what we think of as *understanding*. At this low level of learning, our brain is simply figuring out how to do something, like a motor skill or instant reflex. As infants mature, these reflexes then compound into actions and allow us to do things like walk and talk. Talking, for example, is the combination of several reflexes: our ears pick up different sounds, our brain processes pitch, and our jaw, tongue, lips, and voice box coordinate to produce utterances. All of these reflexes come together and eventually produce the action of talking.

The first thirty-six months of life is pretty much the same for most people. This is not to say that the fetus's uterine environment does not matter, or that the environment that the infant was raised in is irrelevant. Studies demonstrate that the brain is very vulnerable during these early years, and physical problems in utero, neglect, or abuse can actually alter the architecture and functioning of the brain.[5] However, the types of learning skills that are developed at this early stage are simple rather than complex, concrete rather than abstract.

LEARNING HAPPENS IN FITS AND STARTS

When we arrive at the toddler years, the skills children are learning become more complex. For this reason, they learn better when they receive support, or scaffolding. Instructional scaffolding is when a teacher offers support to the learner during their process.[6] This added support is tailored to the learner's needs and helps him or her achieve a deeper level of learning. Learning a new skill such as recognizing letters, sounds, or colors takes *understanding*. A toddler can learn to snap her fingers but does not understand how the sound is produced by two fingers rubbing together. In the toddler years, children begin to form more comprehensive understandings of actions and what they represent. Shown a book that says "B is for bear," they understand that the letter B is not literally a bear but that B represents a bear.

This is where human learning is quite different from animal learning. As far as we know, humans' ability to create and recognize representations is unique, and it is taught and supported by our use of collective knowledge with other humans. With this additional support from other humans, children are able to gain a full understanding of representations by the age of ten and build on that to master abstract understandings. When learners have mastered the skill of recognizing abstract systems, they are able to develop their own principles of thought.

The variation that happens within any one individual is the result of context. The skill scale depicts this as the presence or absence of support. With high levels of support a learner is more likely to achieve optimal understanding. Learners basically continue at a functional level when there are low levels of support.

Every aware teacher knows that with support and the right feedback, a student can reach a deeper understanding of a certain concept or be able to acquire a certain skill. Take this scaffolding away and the same student who could do long division yesterday might not remember how to do it today. So looking at moments when a teacher's support has helped or hindered a child in their learning can be particularly instructive.

Support or scaffolding can be offered in many ways. Your health is a support, as are family, personality, and the weather—the list can go on and on. Support is a part of a learner's context. Dynamic skill theory underscores the importance of the role of context in both shaping and interacting with learning, positively and negatively, and how integration across contexts can take place when learning skills are transferred across domains.

Learning (and understanding) does not always look the way we'd expect. Typically, we use standardized tests to see if a student displays linear growth on the measures assessed by the text. The pattern that emerges from these measurements might suggest that a student has indeed learned. In Figure 2, the wavy line, showing fits and starts, represents general human learning over time, from birth through adulthood. It is also representative of a single student's learning pattern in, for example, multiplication. At first, you might look at this picture of learning and think, "Wow, that is one struggling student!" In actuality what you would have in front of you would be an illustration of a very strong student who is pretty awesome at multiplication. As his understanding progresses, he naturally backslides to incorporate

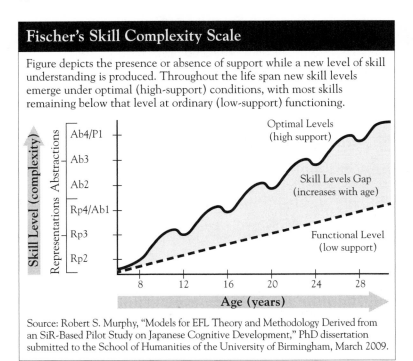

Fischer's Skill Complexity Scale

Figure depicts the presence or absence of support while a new level of skill understanding is produced. Throughout the life span new skill levels emerge under optimal (high-support) conditions, with most skills remaining below that level at ordinary (low-support) functioning.

Source: Robert S. Murphy, "Models for EFL Theory and Methodology Derived from an SiR-Based Pilot Study on Japanese Cognitive Development," PhD dissertation submitted to the School of Humanities of the University of Birmingham, March 2009.

Figure 2

new information that allows his understanding to reach a higher level of complexity.

In education, we have traditionally relied upon linear statistical graphs to measure learning. Such graphs smooth out the results, getting rid of all of those ups and downs so that all we see is a linear line showing growth. However, without the ups and downs, we are not seeing an accurate view of learning.

In fact—and perhaps counterintuitively—the point at which learning is occurring is on the down slope. When the student is on the downward curve, he or she is in the midst of the most intense learning. This actually makes sense in more practical terms. When you are presented with new information that you know nothing about, your stability of understanding drops off because you don't know this new information or skill. If you did, your learning curve would just go straight up. But we are not computers. We don't simply input information into our brains, add it to everything else there, and move on. Instead we take this new information, process it, choose what to accept and reject, and then make sense of it within our individual context. In a sense, during the downslope we reshape the information into something that we understand, and when we do, our understanding grows and that line moves up. So if we want to see what is going on in our brain when we are learning, we should be looking at what is happening during the downward slope.

As we learn new information, we attempt to map that new knowledge onto what we already know, something that requires us to break down preconceived notions—what we thought we knew. Many people can clearly recall this happening somewhere around middle school. I entered middle school with what I thought was a pretty strong academic record, especially when it came to social studies. I knew the names of the fifty states and their capitals; I could recite the national anthem, sing military hymns, and explain the causes of the major U.S. wars.

Then my mind was opened to the concept of perspective. We were learning about the American Revolution and how the United States developed as an independent nation. What was the cause of the American Revolution? The answer was simple: taxation without representation. And somewhere in there we also learned that the thirteen colonies were created to escape religious persecution. Easy enough to remember, and pretty straightforward. But then someone—I now can't recall who— raised the question, "Well, whom are you asking?" The founders of the colonies? Later settlers? Slaves, women, native peoples, landowners? Taking into account perspective, the answer to the question "What was the cause of the American Revolution?" changed each time. My confidence in my ability to answer what I thought were simple questions dropped dramatically.

If I were to draw a graph of my knowledge on the causes of the American Revolution, it would have exponential growth from second through fifth grades, and then you would see a dramatic drop in sixth grade. My understanding of the information I had about the Revolution was being challenged. In order to understand the notion of perspectives, I had to challenge my previous knowledge of the war. It threw everything I previously understood into question. Yet as I transitioned from sixth grade into seventh, my concrete understanding of the causes of wars was replaced by a much more sophisticated and nuanced understanding. It was no longer black and white, with absolute good guys and bad guys or right and wrong decisions. There were perspectives and values, actors weighing priorities and making choices, and others dealing with the consequences and learning from them. The ability to take all these factors into account relies on skills that are undoubtedly more complex than the ability to memorize and regurgitate content.

Just as these dips and dives in our understanding allow brain researchers to better study how we are learning, teachers can see and observe these patterns in their students, and expert teachers are keenly aware of them. When we have started to tear down our previous knowledge, our ability to build new knowledge is often greatly improved by the support we get in doing so.

Remarkably, this same process seen in learning is paralleled in teaching. As we age we develop some teaching skills naturally, but as with any learning, there has to be a dip before there is a brain spurt. With each cluster of spurts we produce a new level of skill and understanding (see Figure 2).

INTERTWINING COGNITION AND EMOTION

Another major breakthrough to come out of brain research in the learning sciences is the discovery that cognition and emotion, always thought to be separate and distinct, are actually interdependent and inextricable. There is no thinking without feeling, and no feeling without thinking. Groundbreaking work by neuroscientists underscores how learning is never cognition alone; it always has an affective dimension.[7] This dimension is yet another aspect of an individual learner's context.

Research has yielded both a theoretical understanding and empirically supported evidence that emotion is a "basic form of decision making" and how "out of the very processes that form the interface between cognition and emotion emerge the origin of creativity."[8] Immordino-Yang and Damasio see this interrelationship as a way to look at the "problematic nature of curriculum," wherein logical thinking skills are traditionally assumed to be distinct from emotional skills. Such a separation of emotion and cognition not only undermines an individual learner's experience, both in the classroom and outside it, but also undermines a teacher's ability to teach. The aware teacher understands that when Johnny, whose parents are going through a divorce, seems increasingly agitated during school or acts out more aggressively with his peers during recess, he is expressing emotional distress. The aware teacher knows that Johnny's schoolwork will more than likely take a dip. Such a teacher would understand that Johnny's ability to cope with the messiness of his feelings will help him stay focused on his math and reading challenges.

Both schools and curricula often separate academic expectations from the emotions that all of us experience daily. One clear way to understand how this effort at separation is contrary to the inextricable neurobiologic connection between emotion and cognition is by looking at how stress can affect learning.[9] Specifically, when the brain perceives stress or assigns a negative valence to a situation, the limbic system (a broad network of brain structures involving the amygdala, hippocampus, cingulate gyrus, and hypothalamus, among others) has one of three responses: fight, flight, or freeze. Depending on the situation and the individual's perception of threat or challenge, he will experience a cognitive impact. Specifically, a positive valence will create a boost in working memory, and a negative valence will create a dramatic decrease in working memory.[10]

Working memory is one of the single most important factors in a student's ability to take in information, manipulate it in his or her mind, and come to an understanding of its significance. When a student's working memory is constrained by negative emotion (such as stress, anger, fear, depression, etc.), he will not be able to think as clearly or remember as well.

LESSONS FROM THE LEARNING BRAIN

As we've seen, teaching and learning involve a lot more than the low-level modeling and skill and drill that come out of the behaviorist, empty-vessel models. And still other inaccuracies arise in the way we think about teaching when the nature of learning is misunderstood or oversimplified. The model of teaching that the current U.S. system of education is based on is a linear one. It suggests that learning progresses like a ladder—that as we age, we graduate from one skill to the next, ideally growing into more and deeper understanding.

The effect of this model is far-reaching. When mothers take their little ones to visit the pediatrician, they are eager to report, "The doctor says he's in the 95th percentile!" Usually parents are looking to see that their baby is developing at the same rate as other babies. The development they look for signals both physical and mental growth. My sister was nearly in tears upon discovering that her son Thomas, three years old at the time, didn't know how to hop. She had gotten a report card from Thomas's preschool teacher indicating the teacher's concern that Thomas could not hop. I tried my best not to laugh, because I knew she was genuinely concerned. She even went so far as to task me with overseeing Thomas's hopping practice during a weekend visit. Although I recognized that technically his inability to hop could be the sign of other developmental delays, I interacted with Thomas all the time and knew he was not developing abnormally. I promised her that this did not mean Thomas would be thirty and unable to hop. More important, I assured her that his current inability to hop well would not hold him back from being a successful adult.

Even though we know that learning happens in the brain and that the brain is a dynamic system, most of us expect it to grow linearly and can't help but compare our own progress to the progress of others. Doing so causes us to miss much of the interesting information about our development in learning—and, it turns out, in teaching.

Learning happens in spurts and the skill of teaching does the same. All of those spurts and dips create patterns of growth we can envision as waves. Over a period of time these waves grow. Each wave represents a new level of understanding. So while linear growth does occur with time, the complexity is in the waves themselves. The rises and falls reveal what is most interesting: the workings of a brain engaged in learning. Like life in general, it's the moments of change and transition, the interplay between failure and success, that truly define who we are.

A traditional linear model loses all of this complexity. These brain growth spurts will nevertheless exist, but the traditional model doesn't allow us to recognize them. And if we don't recognize them, then we can't plan for them. It's not surprising, then, that on the whole we often feel lost and unsuccessful when it comes to the question of how to reform teaching and more generally education. The wrong definition of teaching and an ineffective way of measuring the growth of learners and teachers make it easy, if not inevitable, for us to be led astray.

A dynamic systems approach to learning and teaching is in direct opposition to a step-by-step linear approach. It aligns directly with how our brains function. For researchers, this dynamic view of learning enables empirical support of the notion that there is no such thing as a "good student" who learns by merely receiving information or knowledge. Each student's learning brain develops uniquely and is affected by her personal context, a context that continues to develop and change over time.

Learning is dynamic, just as people are dynamic. The learning brain in all of us can't possibly be an empty vessel that only receives a transmission of knowledge. Learners learn collaboratively with all of the aspects that make up their context, including teachers. Understanding learning in this way gives us another guidepost for reforming our system of education. Instead of basing our changes on unfounded preferences or opting to revert to how it's been done before, we can use this evidence of how we are unique as humans to form the basis for education reform.

Portions of this new understanding about learners are already affecting policy, standards, and the design of curricula. Variation among students is real, and education systems must acknowledge and accept these hugely important differences—and not just because Sally has a high IQ and Johnny has dyslexia. We know that learning is infinitely more complex than that. And as we continue to uncover that complexity, we can't stop there. Equally important and urgent is how this learning brain is connected to the teaching brain.

4

WHY DO WE NEED TEACHERS?

By now the building blocks are in place to define teaching as a natural human skill that is predicated on an interactive experience. This notion, sensible as it may seem, raises an important question: why do we have an entire profession dedicated to something all humans do naturally? And professional teachers reading this book might be angry at the mere suggestion that everyone can (and to some extent does) teach, wary that such a definition carries the potential to reinforce the insulting old adage "Those who can, do; those who can't, teach."

Teaching is not a skill that you either have or don't have. It is not a skill that is stagnant, nor is it a skill that grows and approaches some maximum upper limit that everyone will reach eventually. The teaching skill that is being exhibited in a classroom by an experienced, thoughtful, aware teacher is not at all the same skill exhibited by a toddler, a teenager, or even your average parent. While looking at how children and parents teach can help us see the basic elements and characteristics of teaching, it doesn't allow for a deep understanding of what a professional expert educator does. What makes an expert teacher? This chapter will delve into what distinguishes professional teaching from all other teaching, such as the teaching that is instinctive to a child or the teaching that is done by the most caring and attentive of parents.

WHAT IS TEACHING?

Like learning, teaching is many things at different times and in different people. It's a skill that develops over time and can be honed to the level of expertise by those with natural ability (all of us), training, experience, and great effort and dedication. In fact, we often argue over whether teaching is a trainable skill or an art form. That's the wrong argument. It stems from an inadequate definition that equates teaching with a tool for learning: an individual static variable that delivers information to learners. Redefining teaching as a natural human skill that develops over time, as we've begun to do, means it is both a trainable skill *and* an art form, making it similar to many other skilled professions. Not surprisingly, at its onset skill development in teaching appears very trainable. But upon examining the peak of the skill, I have found that it is not nearly as trainable at that level and requires an expert to demonstrate its potential.

The previous chapter explained the ways in which learning is a cognitive skill that develops over time. So too is teaching. We don't "train" kids to learn. It's true they are trained in their early years to engage in certain behaviors, but as children develop their learning is understood to be more context-dependent. The way they process and understand information becomes more complicated. We don't expect any learner to know everything all at once, nor should we expect that to be possible for a teacher of any type or level. Learners build skills and understanding over time, deepen their knowledge in subject areas, and eventually progress to integrating their knowledge into other contexts. Similarly, teachers build skills that develop more complexity with the passage of time.

The depth and complexity of a teacher's skills depend on the teacher's own experience, effort, and interactions with specific contexts. After earning a master's degree in education (or perhaps attending a summer intensive teacher training workshop), a professional teacher is presumably armed with the basic skills of teaching, including routines, organization, content and skill knowledge, and behavior and time management. But at best this is only the foundation for a skill that will take years of experience, effort, and crafted awareness to develop as an art form that reflects a high level of expert teaching. If the initial training teachers receive—something that comes in many forms and is itself the subject of great controversy—were soil, new teachers would be saplings. Their natural growth process is further facilitated by a variety of internal and external factors that enable them to become trees one day.

The skill of teaching, like learning, originates and develops in the brain. Skills that are fundamentally cognitive call upon the brain as the main contributor. In other words, a skill such as playing the piano is not *purely* a cognitive skill because it also requires physical strength in the fingers. On the other hand, teaching is a set of skills, like learning, that is completely dependent on our brain processing. However, what we haven't fully integrated into our understanding of teaching (or learning for that matter) are the ripple effects that stem from the dynamic systems in which our brains function.

Since we all demonstrate basic teaching skills, the more pertinent question is how do some people become expert teachers. How do they develop a more multilevel approach to teaching? What is the real difference?

I recently interviewed Bob, a high school teacher, about his process of teaching. He described how he is always "pressing students to generate something with more emotional content or belief." Sometimes he would even push a student until he or she cried, fully aware that tears could be an outcome of pushing that way. To the casual observer it might seem as if this teacher lacked sensitivity and was incapable of exhibiting empathy toward his learners. However, Bob explained it quite differently. When he gets to the point where he knows tears are likely, he's already carefully evaluated the student's learning brain and decided that "they are not giving [their all] because they are afraid of something." As he put it, "They're thinking, 'What do the other kids want to hear? What does the teacher want to hear?' Everything but what they want to say, because their opinions have never been valued."

Bob pointed out that safely and gently pushing a student to expose genuine emotion, even if it brings the student to tears, is needed because that's when the wall holding a student back has been broken down and now the student is free. Bob believes that if this wall isn't broken down, he's not doing his job.

Often we associate the experience of breaking down an emotional wall with help or coaxing from someone we trust with other significant life challenges. We have little trouble seeing such an experience as a sign of growth, development, and even acceptance. Bob learned that this type of growth is no less important in the classroom, and as an expert teacher he made it his business to find each student's wall and prompt the student to knock it down.

Though bringing a student to tears certainly would not work for everyone, and I would not suggest that we label it a best practice or promote it in teacher training, it's clear that Bob recognized his strategy as something that worked for him in particular interactions with certain types of learners. More important, it highlights the differences in complexity and multilevel cognitive processing that is specific to humans and our unique teaching capabilities.

Bob is right—kids have walls. And so do teachers. These walls hold us back from achieving our true potential. Some of us learn how to take those walls down, some find ways to climb over them, and others just live behind them and cope. Maybe you remember being a student with a wall. Perhaps a teacher came along—in third, fifth, seventh, or tenth grade—who helped you figure out a way to see that the wall existed and, in time, helped you figure out how to take it down. As a classroom teacher, I always hoped that my students knew my aim was to work with them, helping them become aware of their walls so we could work together on bringing them down.

Malik hung around after class had been let out on the last day of school. Field day and graduation were still to come, but this was the last official academic class. He often lagged behind the others and I would have to nudge him to "get it together" or "get a move on," but this day was different. I could tell from his body language. He was being quiet in a way that was very different from his typical nature. He sheepishly came over to me, handed me a piece of folded paper, and said, "Read it after graduation."

To this day I read Malik's letter when I need to refocus and be reminded of how much teachers influence how students see themselves. These are the lines that stand out most:

> There is a song that goes "We fall down, but we get up." I am telling this to you to thank you because in the beginning of the year I fell down in my school work but thank God for a teacher like you because I got up. I didn't understand why you were so tough on me but I realized that you were tough on me to get up to the level that I was supposed to be on. . . . Thanks for everything and the encouragement. Thanks for helping me reach higher heights.

Zoë was another student who stands out in my memory. I had decided to work at a different school the upcoming year, and my seventh graders weren't exactly happy to learn that I wouldn't be around during their much-anticipated eighth-grade year, when they would finally be seniors, at least in middle school terms.

On the last day of school Zoë handed me a black folder that she'd hand-decorated with glitter hearts. I smiled as I opened the folder and saw a three-page typed letter decorated with more glitter hearts. This was all very Zoë, who was petite, charming, and generally in a chipper mood. But I was surprised by the content of the letter. Zoë began by going through a list of what had happened to her throughout the year and what she had learned. None of it was about the subject we'd been studying, even though her scores on state exams had increased significantly. Instead she wrote about loss, hurt, and happiness.

> I have screwed up countless times this year and disappointed the one person who I should never disappoint—myself. I've finally learned what it's like to believe in yourself and to reap the rewards. You've helped me to understand all of these things and I can't thank you enough for it. . . . You have taught me what it's like to face the real world, that you need to stand up and fight for yourself. I now know that you have to love and admire yourself before you can do anything. The word impossible is no longer in my vocabulary.

Written almost five years apart, these letters reflect my emerging awareness that in order to motivate students, I needed to pay attention to who they were. I had to create a relationship and let them know that I knew something about them. I had to reach them. While there were no doubt other students with whom I was not so successful, I had reached Zoë and Malik—two students who couldn't be more different from each other. They were of different genders, races, cultures, religions, and socioeconomic classes. They came from neighborhoods that were worlds apart, and their personalities were equally distinct. But my teaching brain was beginning to focus on who each individual learner was, to integrate the various dimensions of that child's distinctive constitution, and to draw out the features of the student's background most salient in his or her worldview.

Expert teachers bring this same kind of awareness to their work. Their teaching goes beyond simply showing students they care; by knowing their students more deeply, they will be more effective teachers. Teaching is not a selfless act. It wasn't because of altruism that I came to know and care about Zoë and Malik; rather, it was because I wanted to help them become better learners.

All teachers want to advance their students' learning, and expert teachers understand the work that must be done on the social, emotional, and intellectual levels to make this happen. The expert teacher helps students see themselves and connect with themselves for a sound foundation on which to build up students' content-specific knowledge and skills.

This is what makes expert teachers masters at their craft. They are not counselors, therapists, best friends, or parents. Expert teachers support learning by using scaffolding, motivation, and a continued understanding of the holistic needs of their learners. They do their best to create the ideal mental and physical environment for their students' learning. Parents do this too, but the primary difference between expert teachers and great parents is that expert teachers *plan* to do this through deliberate and analytical means. They explicitly design and carry out this plan in order to achieve precise educational goals. Intentional in design, the application of the plan and any necessary revisions are based on the interactions between the expert teacher and the learner.

DEFINING EXPERT TEACHING

First, the simple definition: expert teachers are systems thinkers.[1] That seems simple enough, no? Well, perhaps not. As previously noted, a dynamic system has many variables. Each variable is constantly interacting with other variables within the system, as well as with those belonging to other systems. They are all affected by and affect one another. Dynamic systems are at once independent and interdependent. It's like the stock market. Individual stocks affect one another, influencing the performance of the stock market. In turn, the market, along with a number of other factors, wields influence on the overall economy. Even a person who doesn't have money invested directly in the stock market will in some way be affected by shifts in the market, via contact with things and people that do have direct connections to the market—say, the butcher where he gets his meat or the store where he buys his sneakers.

How markets and economies work can be confusing to many of us, but not to everyone. Traders on the floor, hedge fund managers, and millions of others understand the market thoroughly. They weren't born with that understanding and in all likelihood they still hadn't developed it after studying it in school; it's generally during their first year of work in that arena that they begin to develop a sense of how the overall system works. Newbie traders learn the parts of the system, are taught how to understand it, and practice within it. Over time, these newbies—and others regularly engaged with the market for one reason or another—come to understand the larger system as a whole. All of those numbers on the electronic ticker tape flying across the bottom of the TV screen or flashing in Times Square actually make sense to a number of people. While many of us may walk past those numbers without a second thought, there are people, corporations, and countries basing decisions on them every day because of the messages they have learned to decode within the financial system.

Expert teachers do a very similar thing with learners. They recognize there are multiple systems in play all at once, and they have the ability to decode those that are directly and indirectly affecting the learner. Expert teachers think, behave, and change in response to the various needs of their students, the classroom environment, and their own personal contexts. While systems thinking has been recognized as one of the necessary skills for twenty-first-century learners, we don't yet recognize it as a skill possessed by expert teachers.

Systems thinking goes beyond the awareness of a learner's academic capacity and social or emotional needs. It allows teachers to tap into cognitive skills that recognize the learner as an independent and complex system. Expert teachers recognize the variables that contribute to the learner's system of understanding and then manage the patterns they create. They keep these patterns in mind in order to make key teaching decisions and in order to adjust their interactions with the learner in a way that will help the student learn more effectively. Redefining teaching in this way, especially expert teaching, leaves open the possibility that less-than-optimal teaching-learning interactions may be a reflection of a teacher's interpretation of a student's learning system just as much as or even more than a reflection of that student's learning system itself.

Peeling back yet another layer of the dynamic system, in order to recognize the student as an individual the expert teacher must also recognize him- or herself as an independent system affecting the student. This recognition of the reciprocal effect between teacher and learner helps the expert teacher to design appropriate interventions that target both the teacher herself and the learner, to meet the needs and goals of the learner.

To sum up, expert teachers recognize the learner as one system, themselves as another, and their interaction with the learner as a third system, which we'll call the teacher-learner system. Expert teachers are able to do this at both micro and macro levels, constructing theories of the learning system for each individual student and for the class as a whole. This multilevel understanding, which comes only with years of practice, is what supports effective teachers as they interact with students individually and at the same time manage to be mindful of the collective learning needs of an entire classroom of students.

WHAT EXPERT TEACHERS DO AND HOW THEY THINK

To examine the nature and extent of the systems thinking approach in expert teaching, I interviewed a group of expert teachers about their process of teaching. The teachers ranged from prekindergarten through graduate-level instructors, and came from rural private schools, urban public schools, and everything in between.

In a description of her teaching process, Leah, a fifth-grade teacher at an elite private school who has taught for over eight years, noted that she "spends a lot of time tailoring [her teaching] to the body [and] with . . . the constellation of the bodies of the class." Her description of the "bodies" in the class as forming a "constellation" highlights her perspective that students are all a part of the system and that they affect each other in creating the atmosphere of the class. She said she was constantly "just zapping away," making rapid, real-time decisions in her classroom.

Leah clearly understood how all of the unexpected events affect the context of her interaction with students: "All of a sudden somebody's nose is bleeding, or a spider is in the room and the [students] are freaking out, and somebody just broke down in tears." Leah was aware of these different events occurring in her classroom, and this awareness triggered her to adjust her lessons in a manner that continuously tended to the students' overall needs. As she put it, she regularly "takes the temperature of the room" and "changes [her] lesson plan."

The external factors that an expert teacher balances within the understanding of the larger teacher-learner system aren't always as explicit as a spider or a nosebleed. Often influences on a student's learning brain happen internally and require a teacher to recognize them on an implicit level. Leah also remarked that students "may be intuitively learning things and then [she] can watch how they're intuitively learning and then that can help [the student] in the lesson." She readily acknowledges using unintentional student feedback to teach more effectively.

Another expert teacher, Caitlin, described being "worried that [she has] preconceived notions about kids." Caitlin, a science teacher in a traditional public high school made up mostly of wealthy, smart, highly driven kids, came from a similar background as her students—white and upper-middle-class—so she sometimes was concerned that her own background affected how she perceived kids. She also recognized that many of her decisions could "decide [the students'] fate." Her concerns revolved around less obvious student characteristics that might drive her decision making. At the same time, she explained, "sometimes [her] gut feeling about students is very correct," suggesting that expert teachers don't avoid incorporating this implicit information into the teacher-learner system. They are able to see how the independent teaching and learning systems interact with the larger teacher-learner system, and are able to arrange or manage the different parts of these systems to help learners reach their goals.

Similar to Bob, Caitlin holds information about individual students in her mind until she needs to use it to benefit the student. One way she does this is to use that information to help build an appropriate culture in her classroom. In one example, Caitlin explained that she took into account information about individual students when considering the group seating in her classroom. She needed to store somewhere in her mind the information that "student X and student Y should never sit together again [because] those two did not work well together, [whereas] these two students, X and Z, work really well together." While it may seem like a small task, Caitlin explained that "gather[ing] the information about kids" and holding it ready for use is quite complex. It requires a micro-level understanding of how the student's learning system (what I call the "learning brain") interacts with all of the other students in the classroom and how the teacher's system (what I refer to as the "teaching brain") can interact with the systems of all the learners in a way that helps support each student's success.

Most significant to Caitlin's work was the culture of her classroom. Each day she did what was necessary to build and maintain a classroom environment in which students are "kind to one another and respected but also held accountable . . . for their actions and for their work and for how much effort they put in." She found it important to "walk into a classroom where [she] want[s] to spend [her] day" and where the students also wanted to spend their day. As Caitlin shared, she spent "a lot of energy as a teacher" doing this "because it's in these very small events that you build the culture of your classroom." In Caitlin's view, the constant interaction of implicit and explicit feedback she received from the students defined the classroom atmosphere and promoted a higher-order educational experience characterized by synchrony (flow) between teacher and student.[2]

As noted above, it is the awareness of these multiple systems and the complexity among them that defines the expert teacher as a systems thinker. Caitlin shared the daily struggle of managing the micro and macro systems within her classroom: "How do you respond when a kid says, 'I need to go to the nurse,' but you feel like the kid might not be truthful? How do you deal with a kid who you know [is] completely disrupting the classroom and you know why the kid is [doing that], you know the kid has a terrible history with alcoholism in the family? So you understand that but you [want to] protect the other kids in the room. I think it's very hard to build that culture, and it actually takes more energy and more thought than one realizes."

TEACHERS HAVE CONTEXT TOO

Teachers who are aware and motivated to fully develop as systems thinkers also understand the level to which their own personal context affects how they interact with individual students and the classroom culture as a whole. Keep in mind that Caitlin's classroom-culture building was specific to *her* values, not a uni-

versal best practice that can be delivered in a five-week teacher training program. It is specific to what Caitlin believes works for her and in turn the culture that she creates to work for her students. It's a reciprocal effect that relies on Caitlin's perspective as the building block for the class.

For Delia, another expert teacher, this meant recognizing that her father's background played a large part in her decision to be a teacher and in how much she valued that lifelong pursuit. Her father "ran away as a teen to England and didn't receive education after the age of 15, so education [has] always been a big part of [her] life." Choking back tears, Delia explained that she frequently reflected back on his experience and knew that what she did in the class each day could "influence the students." She "keeps that in [her] mind as [she] goes through the [teaching] process" each day. Delia also shared that she struggled when first learning to read, as did one of her sisters. This experience directly shaped her interactions with students in the classroom. It gave her empathy, and she "know[s] the process and keep[s] that . . . in the forefront of [her] mind and applies any strategies or techniques and share[s] [her] knowledge." Delia's in-depth attention to a child's reading process is motivated by her unique experiences and family history; she recognizes this and utilizes it to help her students.

Systems thinking also involves encouraging stability in one system by making adjustments to its interconnected systems. This is done by carefully sensing and processing the feedback provided by the interconnected systems. (It's like reading the ticker tape and considering: "What are those numbers telling me about how I need to manage my own stock portfolio today?") Adjustments to the individual system affect the larger system that contains it. In other words, an adjustment that the teacher makes to herself will in turn affect the student and also the interaction as a whole.

As an example, let's consider Caitlin again. She shared how

her current class experiences were affecting her family life. Though she explained that her husband and young daughter were incredibly important to her, she confessed that she had argued with her husband about how hard she worked as a teacher. She admitted that "it's been a source of distress; not so much how much [she] works but how much [she] lets [her] emotional well-being become tied to the school." She shared that after many years of arguments like this she had become more aware of the impact that her work as a teacher had on her life as a wife and mother. Following these moments of clarity (think brain spurt!) she found herself changing how she teaches, seeking to be more efficient in her school work so that she had more time and emotional energy available for her family. Caitlin was constantly engaging in a balancing process in order to maintain the stability of the larger teacher-learner system. She describes these instances as waves, in which she balances these multiple worlds.

As we read these examples it may seem rather unexceptional that teachers would consider the complex system of individual learners, the full class, and themselves (both as teachers and individuals) while engaged in teaching. The events described above are easily overlooked as ordinary but in fact highlight a sophisticated systems thinking paradigm—one that can and should transform the way we think about education as a whole.

While you may have been nodding in agreement as you read through the teacher comments above, it's crucial to recognize that we don't currently have a way of training or measuring the cognitive processes these expert teachers are describing. This is because we don't yet fully understand the unique cognitive, physiological, and neurological processes occurring when we teach. What we can determine with the insights we currently have is that this teaching process, and the underlying skills it encompasses, is in direct conflict with how we typically train and evaluate teachers. We do not speak of teaching as a web of dynamic, complex, interacting systems, and we do not talk about evaluating teachers outside of a linear (and therefore incomplete) model of learning. Instead we train teachers by demanding that they follow linear, learner-based best practices. Even in our most progressive training programs for teachers, training is focused on filling the teacher's brain primarily with knowledge of how students learn. When we evaluate student learning, we attribute a certain amount of the child's knowledge to what a teacher has accomplished. And because of this, there is an enormous gap in teachers' professional development. Or, better said, there is an enormous opportunity.

THINKING SYSTEMICALLY

A systems thinker recognizes the existing parts within a system, how the parts interact, and how they affect one another. Systems thinking provides insights into how a system and its parts will react to changing inputs and allows for interventions that can guide changes and improve outcomes.

Within the classroom there are three core systems: the teacher's system (the teaching brain), the learner's system (the learning brain), and the overarching teaching-learning interaction (the teacher-learner system). Each system is made up of variables that contribute to its individual existence and to the interactive system. An awareness of the system requires being able to receive the intentional and unintentional feedback that the system gives off. An ability to intentionally affect the system goes beyond merely sensing the feedback. It requires processing the feedback, just as a broker does in reading the fast-moving digital ticker tape. Once the expert teacher begins processing this first level of the learner's information, he or she then uses that information to enact a teaching response. This rather simple sense-process-respond interaction allows for the manipulation of these systems. However, one's ability to do this has enormous and varying effects on teaching practices and outcomes.

If Delia's father's experience caused her to pay special attention to struggling learners, while Caitlin's arguments with her husband make her super-efficient at grading homework, imagine their other differences. Those realities represent just one variable for each teacher. What happens when all of these awarenesses come together in one teacher?[3]

An expert teacher recognizes:

- **Each individual part within a system, the multiple systems that exist within a larger system, and that all of these are interacting and affecting each other.** An expert classroom teacher would include him- or herself, each individual learner, the class as a whole, the physical environment, the classroom culture, and many other intersecting parts when describing the dynamic system of his or her classroom. The teacher acknowledges that he or she is an autopoietic system (that is, a system capable of reproducing and maintaining itself) and, as such, requires self-regulation and self-reference. At the same time, the teacher also recognizes the individual learner as a complex system flowing in and out of personal and shared contexts, and that each learner affects the larger system's purpose and the outcomes it produces.

- **All parts must exist in order for the system to run effectively.** A systems-thinking teacher would tailor lessons to include the learner. For example, a teacher might build a series of lessons around skateboarding because that is of particular interest to her students. However, recognizing all of the parts of her class system and tailoring the lessons to leverage her own abilities becomes a part of the process. A former theater major might construct the skateboarding project within a the-

atrical setting in consideration of her own strengths and
interests.

- **Parts within the system must be arranged specifically to carry out their purpose.** A systems-thinking
teacher might map out her yearlong curriculum to follow assumed patterns of child development based on her
experience. Early in the year she may focus on setting
structures and routines, in order to establish the class
culture. This allows for the development of trust between the students and the teacher, which is critical for
tackling more cognitively demanding topics that often
expose learner weaknesses.[4] This explicit arrangement
of the system parts also takes into account the teacher's system; in this example, the teacher may believe
strongly that only in a trusting relationship will students
feel comfortable failing and later learning from those
failures.[5]

- **Feedback is the driving factor affecting the system.** A
systems-thinking teacher understands that while he is
responsible for leading where the students go, he must
constantly adapt his teaching based on the feedback
he receives from students. A systems-thinking teacher
may shift a lesson or a whole unit midstream upon receiving intentional or unintentional student feedback.
He also adapts his teaching based on how he processes
student feedback. This processing is grounded in his
own context, which informs how the teacher will view
any particular situation. Therefore the teacher's shifting
of particular lessons is a result of his internal feedback
mechanism relaying information on how the project
and students are developing. Feedback affects the design
of a lesson, its implementation, and its learning impact. For example, after recognizing the students' acute

awareness of a mass school shooting, the teacher may decide to refocus the upcoming World War II project, shifting from military tactics and armaments to an exploration of the homefront strategy.

- **The importance of using an understanding of systems to predict what will happen and to intentionally manage the system's outcomes.** Systems-thinking teachers utilize this lens to help them predict how the teacher-learner system will develop and move toward its long term goals. They can use this insight to consciously interact with and manipulate its parts in an attempt to attain specific goals. Systems thinkers may choose to ignore certain feedback while reinforcing other feedback, in order to strike a critical balance within the system. For example, in a school where the administration has just recently decided that student cell phone use is prohibited, a systems-thinking teacher would predict the potential for student unrest. Anticipating the negative effect this would have on student engagement and classroom climate, this teacher creates a timely project focused on writing petitions and designing community organizing campaigns in support of cell phone use in schools. This manipulation of the curriculum helps her to reengage the students while achieving the larger content and skill goals related to thematic writing and problem solving. This teacher has recognized various sources of feedback from the system parts, understood their reciprocal effects, and used this to manipulate the system to achieve the intended goals.

Despite such complexity, a linear view of teaching dominates education policy and pedagogy. This narrow scope substantially limits our ability to understand the fundamental mechanics of teaching and the interactions that create the teacher-learner system. However, by examining the characteristics of expert teachers and their teaching brains, we can see how skillful practitioners leverage a systems-thinking approach to guide their classroom interactions.

Teaching is more than we often give it credit for being. It's fundamentally human and interactive—a skill demonstrated by toddlers and parents alike. But it's also a skill, or a set of skills, that can be developed over time and lead to a higher, more sophisticated level—that of the expert teacher. Because of this developmental potential, and because of the context-specific nature that makes each teacher-learner system unique, teachers can continue to learn as they go, and should never stop. All teachers have the capacity to become expert teachers in their context; they can harness their systems-thinking abilities and their awareness of the interdependence of systems (the teaching brain interacting with the learning brain) to create optimal teaching interactions with their students.

These complex factors and layered interactions can be hard to put together and even more challenging to visualize. It's time to shift away from abstractions to better understand what this looks like. The next chapter lays out a revolutionary new framework for understanding the teacher-learner system.

5

TEACHING AS A SYSTEM OF SKILLS

Imagine a student who answers a teacher's question with an angry "I don't know" and then either turns away or stares back defiantly. On one level, a teacher might process this information as the student telling her that he lacks the knowledge to answer the question. At another level, the teacher considers the series of interactions between herself and this student since the beginning of the year: he is young for the class, consistently has had trouble keeping up with the fourth-grade math concepts, and is easily distracted by his peers, often acting out. The student's history of struggling with this subject has made him fear being embarrassed, which causes him to avoid answering questions whenever he is unsure. However, the teacher has also spent considerable time fostering a sense of trust and safety with this student; when his behavior begins to become disruptive, the student and the teacher have a private signal, which the student takes as a cue to quiet down and refocus.

When the teacher considers how to respond to the student's "I don't know," she brings to mind all of this learner's personal context *and* the quality or pattern of their interactions. Based on this context, the student's answer, and his tone, the teacher might choose to push the student to give an answer. However, in doing so, she contextualizes her request by saying, "I know this is a difficult question, but we can work through it together." The student accepts this revised situation based on his existing trust and experience with the teacher. He decides to take a risk and engage in a discussion. It is in this interaction that learning and teaching take place.

The act of successful teaching occurs in a large, encompassing system, in which the teaching brain and the learning brain are dynamically interacting. Indeed, teaching and learning are part of the same dynamic process in which information flows and feeds back between the teacher and learner. Within this larger system, several levels of interaction occur (see Figure 3):

1. The teacher senses student-centered information.

2. The teacher **processes** the information he or she has pulled in that is relevant to how the student learns.

3. The teacher then utilizes this processed information to **respond**, offering the student support to improve the depth of his or her learning.

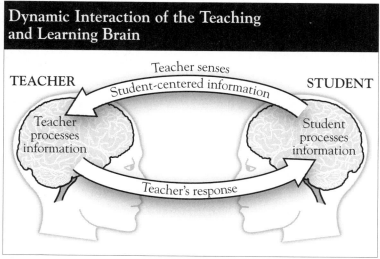

Figure 3

The student takes in the teacher's response, processes that feedback in his or her learning brain, and then the process begins all over again until the learner has reached their learning goal. These steps, while presented here as abstract, are very tangible experiences in teaching. Expert teachers are constantly refining how and why they do what they do based on the input from the three sources.

THE NERVOUS SYSTEM AS A METAPHOR FOR THE TEACHING SYSTEM

The human nervous system provides a solid analogy for how this complex system of teaching is organized. The nervous system has three parts: the afferent or sensing limb, the central processing unit (spinal cord and brain), and the efferent or responding limb (motor functions). Essentially, our nervous system is set up to sense our environment, process signals from the brain and the body, and respond to those signals with thoughts and actions. Some of these responses are automatic (e.g., body language), while some are more consciously in our control.

Let's take a simple example of how the nervous system operates at the most basic level. Your body senses the temperature of the surrounding air and then processes that information by shooting a signal up the spinal cord: *It's cold!* Your body responds with an automatic reflex—you begin to shiver. Beyond that, you might do some additional cognitive processing and decide that, in light of the shivering, you need to put on a coat or go inside.

This is an oversimplified version of what happens very quickly in the nervous system as a whole. But this trio of physical-emotional-cognitive actions—sensing, processing, and responding—happens all the time and characterizes the way that teachers begin to strategize and then carry out their teaching. As we've seen, teaching is not simply a unidirectional process based on how the teacher downloads information into the learner. Rather, the process is marked by a complicated web of interactions entered into by both the teacher and the learner.

At the most basic level, which I call **spinal cord teaching**, teachers begin organizing their strategies by first pulling in information from around them (sensing) and then responding reflexively, with very little if any deeper processing. Reflexive teaching is the sort that is not very cognitively demanding, and thus can be easily trained and scripted.

At the next level, which I call **student-centered teaching**, the sensing-processing-responding arc is more complex. Teachers tune into students, their developmental levels, their context, and who they are. They also begin to process how this information affects their teaching practice: *Based on this information, how should I structure a lesson? Who are these students as individuals? What are my goals of instruction, and how does my knowledge of students bear on them? How should I develop or present my curriculum in light of what I know about these learners?*

When the student-centered teacher responds, he is working at multiple levels: sensing or pulling in his knowledge about the subject, understanding or processing the students' needs, and in his response aligning these needs with the curricular goals. The nature of the response is key because that is where we can actually witness the level of cognitive processing engaged in by our teaching brain.

At the highest level of teaching, a teacher is not only aware of the basic concrete needs of the learner and the context of the environment but also is highly aware of his role in this interaction. When a teacher is aware of the complexity of the interaction between himself and the learner; takes in information (both current and background) on the student, the environment, and himself; and understands that the interaction is dynamic, not static, the sensing-processing-responding act shifts and teaching becomes more cognitively demanding. The teacher is using more of his brain to teach.

SPINAL CORD TEACHING

Spinal cord teaching is the most basic form of teaching. It can be found in humans and in animals, as we have seen in earlier chapters. It is also the type of teaching most common among young children in peer-to-peer relationships, and in parent-child interactions.

In the act of teaching, teachers activate their sensing limb to gather information that will advise their response. Teachers sense two types of information from their learners: intentional feedback and unintentional feedback. Intentional feedback is when a learner actively gives feedback to the teacher. Unintentional feedback is more subtle. The learner does not actively or purposefully provide direct feedback to the teacher regarding their learning ability, preferences, or achievement, yet from the situation the teacher is able to gather feedback from the learner, and their context, that will influence their response.

At the level of spinal cord teaching, teachers respond directly to the observed learner's behavior or respond based on their own teacher-driven goal. For example, a second-grade teacher has students copy simple spelling words three times each, trusting that this method is the best way for such young learners to familiarize themselves with the correct spelling. Very direct, very simple.

Spinal cord teaching is hardwired and reflexive, requiring little higher-order thinking and processing. Let's think back to the pied babbler from chapter 1. Research has shown that these birds do in fact have the ability to teach. The mother bird trains its young to learn a bird call that at first indicates when it is feeding time. Each time the mother sings the same tune and the baby birds are given their food. In true Pavlovian form, over time the young birds equate the call with food. This call is then used for several purposes, such as luring the young away from danger. It is hardwired from a young age—the mother bird has indeed taught her young ones the call, utilizing the common practice of reward and repetition.

The mother bird is not concerned with who each offspring is as an individual or how each of them learns best. She is doing the same call for every bird. She does not consult the young bird to see if it prefers a different call or no call at all. The feeding call has a goal that is set forth by the mother and carried out, likely with the best intentions for the young birds at hand (namely their survival), but not based on anything more. The mother bird recognizes the knowledge gap and acts to close that gap through a demonstration of how to gain food by responding to the feeding call.

The premise of spinal cord teaching is that it is based neither on the preferences of the teacher nor on those of the student. Spinal cord teaching is based on very specific expectations of outcome—think passing or failing a standardized test. Following a simple input-output model, a spinal cord teaching response is hardwired. The teacher responds to an observed behavior in the same way every time, regardless of the learner. This type of modeling or demonstration does not necessarily suggest that the response is accompanied by an explanation from the teacher. It also does not involve complex processing by the teacher that includes a consideration of why a given teaching response has been chosen.

Similar spinal cord teaching practices take place in our classrooms today. Chapter 2 related the experience of a young charter school teacher whose school principal interrupted her math lesson to correct her on how she taught students to celebrate. "It's two claps and then a sizzle," the principal stated, reinforcing the celebratory behavior required of all teachers in all schools within this system of charter schools. Teachers, students, and even administrators all follow the same celebratory behavior, which was not created or modified based on any preferences or feedback from those expected to participate in this practice. When a behavior is observed that requires a celebration, the response is two claps and then a sizzle—period. There is no cognitive processing necessary, just a reflexive response. Teachers in this network are taught this behavior in a fast-track summer training program, and students are in turn taught it upon entering the school. Within a particular school, even across schools within this chain, anyone could enact this call and children and grownups will understand how to react. They have all been taught how to celebrate.

Another way to grasp spinal cord teaching is by considering the two ways that (from the perspective of most lay people) computers obtain data. Either a computer has been programmed to pull information from a data source or the information has been uploaded manually by an outside source. (It's also possible for computers to generate new information, but that is a more complicated discussion. For our purposes, let's stick with this early, simplistic version of computer processing.) Computers are able to train and demonstrate in an effort to close a knowledge gap. If we were to ignore for the moment that a computer does not spontaneously (or willingly) sense and retrieve information, then we could add it to the list of spinal cord teachers.

Like computers, teachers in the position of *only* responding to a learner's performance on particular tasks do not require a complex understanding of the student's learning brain. Nor need they be aware of their own personal context. In this model of teaching, learning of a particular sort (generally the accumulation of information) can and often does happen because the teacher is positioned solely to service the learner. This type of teaching occurs all the time and is not to be scoffed at.

Often we use spinal cord teaching as a mechanism of survival. My nephew Thomas, who is now seven, takes his family roles very seriously. He learned how to make oatmeal by listening to and following instructions from my sister and mother. In turn, Thomas has passed along to his younger brother, Jordan, and his cousin James many instructions for how to make oatmeal, including telling them not to get too close to the hot stove. Every time the boys get close to the burner, he delivers the warning with the same tune that was often used with him, a quick steady rhythmic sound that is much like tapping: "Hot, hot, hot, hot."

It's really just instinct for Thomas, and it's becoming the same for Jordan (now four) and even James (now two). If anyone goes anywhere near the stove, James will say, "Hot, hot, hot, hot!" James can't demonstrate why the stove is hot or how not to touch it, but his instinct is to say "hot" whenever someone gets close to the stove. Spinal cord teaching helps us to pass on fundamental and basic skills, abilities that in many cases are important to our very survival.

STUDENT-CENTERED TEACHING

As teachers become more aware of their learners, both their teaching and their interaction with the learners become more nuanced. Returning to the metaphor of the nervous system, teachers are sensing two types of information from learners: intentional feedback and unintentional feedback. This I call *teaching brain processing*, which is distinct from purely reflexive *spinal cord processing*. Student-centered intentional feedback may come in the form of a direct question in which the learner shares a problem he is encountering or a lack of understanding. Unintentional student-centered feedback might include a student's stomach grumbling, the first snowfall of the year, or a typically upbeat student suddenly becoming quiet and withdrawn.

Eva's dad, introduced in chapter 1, is an instructive example of teaching brain processing. Eva did not explicitly ask her dad to teach her how to close a button on her coat, but she did intentionally express that she could not close her coat. Eva's dad heard her complaint and also picked up on Eva's general frustration. He respected her existing knowledge by using the same language that she had developed: the buttons on this coat did not have a "nose" and were not able to "sneeze," which was Eva's way of describing the sound a snap makes as it clicks into place. Eva's dad considered that the frustration stemmed from his daughter wanting to learn how to close this new coat; importantly, however, he gave her the right to choose whether she wanted to learn. The response is based on Eva—not just specifically who she is, but who she is as a learner within her context.

Eva is inquisitive, so it is likely that she wanted to learn. Yet her father had to consider that they were exiting the car and that he was dropping her off at school. Eva was often very anxious about being on time for school, but equally anxious about not knowing how to do something. Additionally, Eva struggled with fine motor skills, which is why her mother typically dressed her in a coat with snaps rather than buttons.

Eva's father used his teaching brain to process both the intentional and unintentional feedback he sensed from Eva's learning brain. What matters is not simply a consideration of the biological factors that affect how Eva learns; Eva's learning is context dependent, and the variables within it are infinite. Her dad's student-centered response required multifaceted brain processing, not just a spinal cord reaction.

With unintentional feedback, the teacher does more intuiting and analyzing about a student's needs. In the classroom, for example, a teacher would take note of a child's body language, mood, and overall response to a lesson. Is the student in the back paying attention? Why does she look so distracted and tired? This unintentional feedback is both tangible and intangible.

THE CASE OF LOGAN: A STUDENT-CENTERED APPROACH

When Logan first entered my seventh-grade class it was hard to see him—quite literally! His hair was shoulder length and always covered his face. Maybe if I had grown up listening to more Led Zeppelin or Rolling Stones instead of Latin freestyle, I might be able to offer an apt visual comparison to some rock icon— but my knowledge is severely lacking in that regard. Logan often wore black T-shirts with one or another rock band logo on the front. Accompanying the curtain of hair shielding his face was a hunched-over posture, generally cloaked by a sweat jacket and perpetually crossed arms. Logan didn't speak unless he was forced to, and even then he tried not to. When he spoke he either mumbled or blurted out a rather disagreeable statement, such as "Why do we always have to do group work?"

These intangible pieces of feedback weren't the only information I had about Logan. His scores on standardized state tests in both math and English language arts were excellent. He didn't seem to have trouble with long- or short-term memory, both of which play a major role in performance on standardized tests. I can't say the same for his in-class scores on tests and projects. He rarely did homework, and when he did, it was the bare minimum. This was the case in every subject area. His parents described this as their main source of frustration. I met both of Logan's parents at back-to-school night early in the school year. They were eager to tell me, in their own words, who Logan was as a student.

I was used to parents telling me who their child was, so this wasn't a surprise. I was so used to it that I did my best not to let parents' comments completely change the theory I had already formed about the child. After all, a parent's lens is just that, a parent's lens, which is (and should be) different than my lens as a humanities teacher. Logan's parents came to the meeting dressed in business suits, and I can recall them explaining that they didn't know where Logan got his lack of motivation from, because both of them had been dedicated students up through graduate school and were quite successful in their current positions. They felt strongly that Logan was capable of doing the work and used themselves and his standardized test scores as evidence. They also had specific homework hours at home for Logan and had enrolled him in test prep courses in addition to hiring a college counselor. "We never had this problem as students," they reiterated. "We're sure Logan is just being a slacker and could do the work if he tried."

This was not the type of parent meeting I was used to. Here I had parents apologizing for their child and telling me what he could be (but wasn't) doing. They didn't come close to blaming his lackluster performance on something I had done. Logan's parents thought he didn't care about school. "We've tried everything; he just doesn't care. He won't listen to us, just slams doors and fights with us at every step." From my experience teaching adolescents, I've found that if they're still slamming doors and fighting with you, it means they care. Doing the opposite of what they're asked to do is a clear message, and it requires effort.

I constantly revisited the process of trying to build my awareness of Logan as a learner. It was typical trial and error. What was Logan's theory of mind? Did he just love rock bands and long hair? Was he high most of the time and trying to cover it up? Did he think this image would attract attention or deflect it? And whose attention was he trying to gain or avoid—mine or his peers'? Based on his interactions with other students, the theory I formed for Logan around his appearance and dress was that he was hiding. He seemed to retreat and hunch over more whenever anyone, friend or crush, attempted to look beyond the layers— if, say, someone tried to brush the hair away from his face, he'd swipe it back into place. He seemed to speak only when he lost control of his emotions, which seemed to happen when he disagreed with a task but couldn't easily avoid doing it. Though he did his best to hide, I sensed that Logan's emotions played a large part in his choices. It didn't seem coincidental that Logan appeared to be the exact opposite of his parents. Logan often alluded to not caring about the difference between what they expected from him and what he produced, yet he made a point to recognize and verbalize the distinction. Each time a teacher told Logan he wasn't living up to his potential, he seemed unconcerned and even somewhat satisfied.

I had to determine the best way to respond given what I had to go on. Sometimes I went the traditional route and called on him during full-class discussion (I know that can be horrible, but a teacher has to pick and choose when this move is an appropriate one). However, the bulk of our class time wasn't spent in full-class discussion. And even though I am an advocate for group work, it can be a perfect way for a student to hide. Since I knew Logan was able and had the expected content and skill knowledge, I'd often challenge his instinct to hide during group time. Sometimes this meant putting him in a group of other students who I knew were also reluctant talkers. Sometimes I'd group Logan with students who I knew also struggled to get their work done, with the belief that this would force at least some of them to step up to the plate rather than sitting back and letting someone else produce all the work. During an extended project on popular banned books, I assigned Logan to a book club group with other boys who also described themselves as outcasts, and they read *The Outsiders*. In our book clubs students were taught to have evidence-based conversations in which they shared their perspective; they took turns introducing a topic derived from their reading and created questions for discussion. They were required to vary their sentence starters based on cognitive complexity, which was measured on a scale from 1 to 3. For example, questions about names, places, and dates carried level 1 complexity, while questions about cause and effect rose to level 3. We practiced how to balance talking time with listening time. I made a sincere effort to make our book clubs true to adult book clubs, while adding in necessary scaffolding to support my adolescent learners.

Try as I might, I always felt like in Logan's case I hadn't quite gotten the right *theory of the learner's brain* (ToLB) for him. (I'll discuss ToLB in more detail in the following chapter, but in short it is the theory that a teacher forms in regard to how a student learns.) Logan still struggled to meet the school's expectations, and though I believed he was brilliant, I had only my perspective to go on, rather than acceptable data. I certainly tried common best practices (homework help, after-school help, evaluation for occupational therapy, parent intervention, rewards and consequences, and so on). But here I want to focus on the strategies that were more cognitively demanding, such as forming a complete ToLB. Considering that most of the feedback I had on Logan was unintentional, it's not entirely surprising that it was such a struggle to develop a solid theory about his learning process, but it sure was frustrating.

Through it all I had no idea if Logan dismissively grouped me into the category of other adults he thought of as frauds and pretty much useless. I wondered if he knew that I was really trying to figure him out and create a space for him to express who he was as a learner. With our last project I made a bold move. In our reading and writing workshop I had decided, with another humanities teacher on my team, to do a full-class read-aloud of Meg Rosoff's novel *How I Live Now*. It has since become a movie, but at the time it was fairly unknown and for a seventh-grade class quite controversial. I considered the project to be a mixed-methods approach, combining both quantitative and qualitative research. First, students read through the book using what I called the scientific method for reading—noticing the typical book topics (plot, character, setting, theme) as students became scientists of literature. They went through the traditional process of the scientific method: pose a question, complete background research, construct a hypothesis, test the hypothesis, analyze the data, and report the results. Their question had to derive from or be inspired by the book. Their data therefore were also gathered primarily (though not solely) from the book. Their questions changed as we moved through the book, as their interests changed, or as they proved their hypothesis wrong.

For Logan, I had hoped this process of asking questions and finding evidence to support his hypothesis would offer the opportunity to feel safe speaking out rather than using silence as his preferred method for fighting back. It didn't work as I had hoped. That he liked the story line and the characters was evident from the way he placed himself front and center during read-aloud time (his usual spot was behind the supply cabinet). When we had quick "turn and talks" with peers, he actually spoke. But Logan continued to struggle to complete his written work, which meant that I had to rely mostly on unintentional feedback to determine my next move and to assess his learning.

The last trick up my sleeve was to offer Logan the opportunity to co-create and role-play. After reading the book we focused more heavily on our writing workshop project. Students were asked to take on the perspective of a character in the book and rewrite five major elements of the book from that character's lens. Writing from the perspective of another is a pretty standard request from an ELA teacher. To further layer the perspectives, each student was given both a character from whose perspective to write and a character to whom they would retell the story in the form of letters. This drove home the idea that the letter you'd write to your beloved mother who thinks you're an angel would be quite different from the letter you'd write to your best friend or estranged lover.

For Logan I chose the perspective of Isaac, a character who

had at most two lines of dialogue throughout the book but was looked upon in a positive light. Everything else known about this character was based on how the heroine perceived his unintentional feedback. Logan had full license to create the story of Isaac. He was assigned to write to Isaac's mother, who, though extremely bright and successful, was unable to help Isaac in his time of need. I thought of this assignment as offering freedom to live through a character—the equivalent of wanting advice from a friend but being too embarrassed to say it's for you, so instead you say, "I have this friend who . . ."

I hoped this would finally be the crack that would let in some light. I'd considered the ToLB that I had painstakingly developed and modified for Logan. I understood his motivation to be emotionally connected to his disconnect with his parents. I thought that facilitating a way for him to release and explore this dynamic would shift his motivation toward investing in his school work. Initially his one-on-one conference conversations with me improved, as did his participation in small groups. But he was completing less work overall, and after one or two letters he really floundered and we hit a wall again.

Logan's story doesn't have an ending typical of the heroic-teachers-saving-struggling-students stories found in movies such as *Freedom Writers*, *Waiting for Superman*, or *Won't Back Down*. But my choice to showcase this anything-but-ideal example is quite intentional. This discussion is about how teachers practice student-centered teaching—not whether the end result matches common measures of student success. The truth is that, as with most things, there are great successes and troubling failures. But the process matters. The more deeply we understand the process, the easier it becomes to tip the balance toward great successes.

In the end all of my efforts to address Logan's needs did not create what I would define as synergy. We did not come together as a pair, and though there were some incremental steps forward, neither did we find the best way to interact so that he could reach his optimal learning level. However, by all measures this was an excellent example of a student-centered approach and was just as important to the development of my teaching brain as any hard-won success would have been.

THE LIMITATIONS OF STUDENTS AT THE CENTER

Logan's case highlights how complicated a student-centered response can be. I had been trying diligently to tune in to him and define a ToLB for him, but I was still off. When teachers form a correct ToLB, they are more likely to have successful teaching interactions because the student to whom they are tailoring their teaching is actually the student in front of them. We know that these interactions are fundamental to learning and development.[1]

Many education reformers and researchers have therefore focused their attention on developing teaching models and evaluation methods that support or even demand a student-centered approach to teaching. We saw this highlighted in chapter 2, where molecular biologist John Medina's theories on how the brain learns were translated into concrete changes to a lesson and to the school day. (See Table 1, page 38.)

While a student-centered approach is a noble goal, it also misses the reality that teaching is an interaction and that the impact of another human—the teacher—on this equation is inescapable. Many other common brain-based curriculum reforms focus specifically on the learning brain and recommend teaching with only the student in mind. Researchers for whom I have deep respect have aimed their findings toward the development of curriculum goals, methods, materials, and assessments grounded in evidence of brain networks.[2] Frameworks such as the Universal Design for Learning (UDL) have focused on integrating flexibility into curriculum design and the learning environment, so as to better target the inevitable variability in any classroom. UDL also addresses ways teachers can manage and address the dynamic interdependence between affect and cognition. However, it remains a student-centered teaching approach, based primarily upon the analysis of the learner's brain.[3]

Since a student-centered teaching response is, by design, focused only on the learner, it ignores the impact of the teacher in the construction of any theory formed about the learner. Though the teacher is actively sensing student-based feedback, this information is not processed independently nor without perspective. Instead it merges with the teacher's personal context and is viewed through the teacher's personal lens.

Teaching methods designed around only the student and the learning brain attempt to limit teaching responses to the student's learning brain. However, that is an unattainable goal. We will never know definitively what is happening in a learner's brain. It's the equivalent of saying there are no indisputable truths. In science we do not say something is false; instead we say we have not been able to prove the null hypothesis. The same is true when it comes to figuring out learners. We can only form theories of the learner and then set out to disprove our null hypothesis, but within that we must account for the lens of the teacher. You might come up with a correct theory of your learner by chance—consider the rule of averages—but if you believe that learners are all individuals, then forming a nuanced theory of each individual's learning brain requires being aware of the multiple factors that shape and influence it.

To create a comprehensive teaching response that acknowledges the impact of both the teacher and the student on the interaction, teachers must process the information that enables them to reveal the multiple awarenesses that impact their teaching. (The five awarenesses include awareness of learner, interaction, self as a teacher, teaching practice, and context.) In the end, reflexive and student-centered teaching can get us only so far. Without sufficient focus on the role of processing—processing

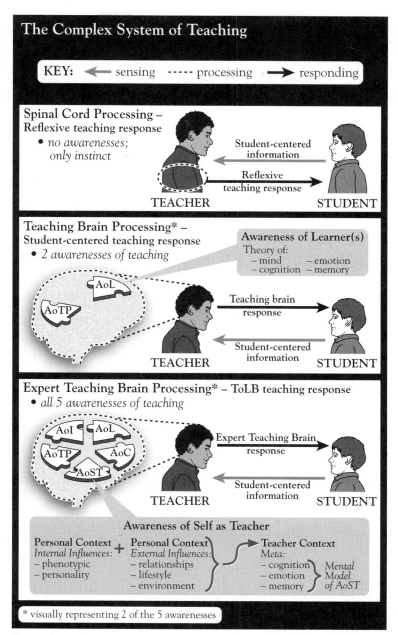

Figure 4

not only of the factors affecting the learner's brain but also those affecting the teacher's brain—the entire interaction is stunted. The teacher must be acknowledged as part of the teacher-learner system; only then can the best learning be co-created. Expert teachers sense, process, and respond to both the learning brain and the teaching brain.

PART THREE

THE TEACHING BRAIN

6

BECOMING AN EXPERT TEACHER

Spinal cord (instinctive) teaching and student-centered teaching both have important roles—for learners as well as teachers. However, in order to evolve into an expert teacher, a certain cognitive and psychological development needs to occur, a development that doesn't simply happen naturally with age. It takes experience, attention to teaching practice, and awareness about the nature of the teacher-learner interaction.

In this chapter, we will peel back the outer layers of what expertise in teaching is, and talk about what it isn't. We will see the full capacity of the human teaching brain at work.

THE FIVE AWARENESSES OF THE TEACHING BRAIN

In order to teach, one must process information both as a teacher and as a learner. There is not some wholly separate brain that is our teaching brain. Instead, identifying the teaching brain is an attempt to highlight several patterns within brain processing that characterize the complexity of how expert teaching develops—from basic reflexive processing through forming multiple abstract systems of teaching.

The key to expert teaching is developing what I call a comprehensive theory of the learner's brain (ToLB). There are five awarenesses (as depicted in Figure 5) that constitute an overall ToLB.

Awareness of Learner

As teachers process student-centered information, they begin to form a theory of mind, theory of cognition, theory of memory, and theory of emotion of the learner. Theory of mind (ToM) was discussed more extensively in chapter 1, but in short, it is what we perceive to be the reason behind someone's behavior. It may not seem like a major feat to consider the reasoning behind someone's action, but I would argue that it is. Imagine being a store manager trying to decipher why multiple employees continually forget to restock a certain product; or being a mother of four children, each of whom seems to require a different explanation of the proper way to complete house chores; or being the math teacher of thirty students trying to learn geometry. But no matter the circumstance, a theory is just that—a theory. It is not a simple exercise to attribute a mental state to another being, let alone to ourselves. Think of how much trouble you've gotten into with a wrong guess. My husband's recent drive to be more "green" did not mean that he would like to go on a surprise freegan dumpster dive for his birthday. Let's just say the ToM I'd developed for him that year was a bit off. (The next year we went clam digging.)

I believe that teachers do a lot more than form a ToM for their learner. While some opt to group many components under ToM, the presence and importance of each contributing element merits distinction. A theory of cognition is an ability to consider a person's understanding (based primarily on her or his perception and motor control). There are many theories of cognitive development (Piaget, Vygotsky, Dewey, Fischer) and though they are different, they are all models for how we might uncover an individual's process of knowledge formation. Allen Newell called for one unified theory of cognition (UTC), which would allow us to create rigorous models of cognition supported by experimental data. There are no hard, indisputable data on whether teachers form a theory of cognition for their learners, nor can we fully measure the quality of any theory of cognition teachers do produce. My aim is to offer a framework for how we can begin to describe the cognitive processing of teachers, which up until now we have only glossed over by labeling it "complex."

So what would a teacher's theory of cognition look like? A teacher might form a theory of the learner's cognition by asking questions like: "Is it better to have David produce an essay to explain his understanding of the book, or does a book club format better support his thinking?" "Why is Melissa asking me about Rome when we're learning about World War II?" This brain processing is never-ending. A teacher creates a theory of cognition for each student and for the class collectively, and calls upon these theories throughout the entire process of teaching—during planning, interacting, and reflecting.

The same is true for forming a theory of memory. Teachers often have to make choices based on whether they believe their students have remembered information and whether that information is stored in the students' long- or short-term memory. Not long ago I was asked to review a survey designed to uncover how comfortable teachers are with students' wrong answers. I was asked to rate both the clarity of the survey question as well as whether the question was relevant and answerable. One of the questions read, "If a student gives an incorrect response, how soon should a teacher provide a correct answer?" Well, it depends. I may never give my student the correct answer, because the answer is not always the point; instead, the goal might be to help the student build her process of cognition. However, in order to do that I would consider the student's ability to remember what she has learned and evaluate the appropriate amount of time necessary for her to recall that information. It might be a matter of seconds, days, or even months. The amount of time required for the student to recall information depends on many factors. Is she recalling a fact, a skill, a process? Does it require hints? Do I have to scaffold with a demonstration or explanation? The answer to the survey question is that the teacher should first form a theory of the learner's memory so that the teacher could respond appropriately.

Another consideration is, of course, the learner's emotional state. As a teacher, I form a theory of emotion for my learners. I consider how they might feel if I were to ask a question, introduce a topic, or partner them with a particular student. Think about how often your emotions drive your reactions. Emotion is tied to motivation and to the ability to cope with stress and negative feelings; emotion impacts working memory and the ability to process information. All of these emotional sequelae are critical to how a person learns—both over time and at any given time.

During my first year as a doctoral student at Harvard, I was sitting in a lecture with my cohort of peers as a renowned professor spoke about student achievement. Projected on the screen in the large lecture hall were the typical tables and diagrams that usually spelled out the same message: poor students of color do worse than their affluent white peers. Having been a teacher in New York City's public school system, I was so used to these statistics that normally I barely paid attention to them anymore. I had even helped collect such stats, serving as the data inquiry lead on several school committees.

But as a new doctoral student, I was excited to finally learn how these more complicated figures were really created. What was the research question? How did they decide on the measure? How were samples chosen? What data analysis technique was used? I wanted the real story behind the numbers; I wanted to understand in depth what it meant when my principal handed me a stack of student data and said, "We have to get our scores up by 10 percent." But these slides were different. Rather than the categories I was used to seeing for demographics (American Indian or Alaskan native, Asian/Pacific Islander, black/African American, Hispanic American, white/Caucasian) there was a column specifically labeled "Puerto Rican," and the study had been done in New York.

When I saw that small new detail, something happened that I never could have anticipated. Suddenly I felt that the professor was talking about me. As he went through the subsequent slides, I was still stuck on the first. *He's talking about me and my family* was the refrain ringing in my head.

The slides were no longer information about statistical averages and standards. The median income slide was now a slide about how much money my parents made, and according to him I was poor. The hard numbers didn't matter. Saying or implying my family was poor felt insulting. The next slide showed level of parental education. Listening to my peers, I heard enormous judgment about the people and the circumstances indicated on the slides. My internal conversation became heated, even confrontational. *Of course you'd look at it in this light,* I thought. *Most of you come from affluent white homes that aren't first generation!* Having spent years learning to be proud of the nonmonetary riches I'd grown up with, I found that suddenly being personally cast as little more than a statistic was hurtful and dehumanizing.

The lecture went on and on, and finally I had to step out of the class when it became obvious that my emotions had eliminated any possibility of my processing what was actually on the slides, or the purpose of the lesson at all. This, of course, was not a typical reaction. I hadn't made it that far in academia by storming out anytime my emotions welled up, and I surely couldn't have survived being a classroom teacher with that sort of MO. But this experience struck a particularly sensitive chord, and because of that I had trouble learning the material.

Imagine this scenario from a teacher's perspective. (Since you have ToM, you can do this!) In all likelihood the professor did not intend to elicit this reaction. It would have been nearly impossible for him to do so, since he knew nothing of my background. I can't even say that I could have predicted my emotional response had he asked me ahead of time. I was used to identifying with the Hispanic category, but such a large aggregated classification offered a comfortable anonymity, so it had never felt personal. But had he known more about his students and their learning brain contexts (something that's arguably a stretch for someone teaching a large lecture class), he might have developed a theory of emotion for me and made different choices. I know that had my emotions not been so negatively charged that day, I wouldn't have shut down my learning.

Whether they are aware of it or not, teachers come to the teacher-learner interaction with a developed sense of who that student is, how they think, what constraints might be in place in terms of memory skills, and how able they are to manage their own emotions. That's not to say that this sense of a student is always fully or accurately developed, but nonetheless it is present.

These multiple theories of the student as a whole are what a teacher uses to create a comprehensive "mind map" of the student's learning brain: an overall awareness of the learner (AoL).

Awareness of Interaction

In a recent study I completed, my interviews with teachers suggested they were aware of at least four categories of interaction with a student: connection, collaboration, mutual effects, and synergy.[1]

Connection

This type of awareness of interaction (AoI) refers to instances when teachers describe the tight relationship or deep connections they have with students beyond the necessary interchanges involved in academic content. It appears that at its most basic level, teachers realized that they form relational bonds with their students that are essential to the learning and teaching that happens in their classrooms. In describing whether there is a distinction between her "real self" versus her "teacher self," Meredith, an elementary school teacher, said:

> It's back to that connection piece. . . . I want children to know that when they walk in the door, I'm really authentically happy to see them. And it matters that they came to school, and it matters that we're going to spend this time together. That they know that when I speak to them, I'm going to do it respectfully, and that I have the same expectation of them for each other, and that it's not just about, I guess, that I'm setting rules for how people interact with each other. It's about knowing each other, and why that is important.

For Meredith it was important she and her students have a relationship where they understood each other authentically in order to connect on a personal level. This personal connection was at the center of her identity as a teacher and foundational to her processes and the decisions she made in her teaching.

A pre-K teacher named Andrea spoke of a time when her personal connection with students allowed both her and a student to share feelings that were critical to their mutual success. A few years earlier, Andrea had had to take time off and travel back home when her father passed away.

> When I came back I told the kids that I was really sad, and I said, "I might cry, and I want you to know that that's okay if you see me crying." Basically [students] then took on this empathetic role to take care

of me. Not in that sense of that it was like I was putting that responsibility on them, but it was like, "All right, okay. So Andrea's feeling a little sad—what are we all going to do? How are we going to step up our game?"

Then one little girl just said to me, "You know"—and it wasn't, like, egocentric or anything—she was just like, "You've been gone a lot. Don't do that anymore."

[So I replied,] "Okay. All right. So I'm not going to go."

Andrea concluded, "For me the value of being a really close community is that she even felt that she could say that to me."

Andrea's example highlights a teacher's ability to both express her feelings and recognize a student's feelings due to the tight relationship that they share. Her example demonstrates the awareness of relationship and connection that allows teachers to be authentic human beings while also empowering their students to express the feelings that affect their learning. As another elementary teacher, Laura, described it: "When a teacher really knows a kid and can connect to that kid, then, you know, that's going to be the surest path to academic success." Teachers who identified the relationship or connection aspect of interactions framed it as necessary for their success and their students' success.

Collaboration

This type of AoI refers to instances of teachers and learners sharing knowledge, work, and responsibility. Teachers who defined the teacher-learner interaction in these terms talked about working with students to achieve a common goal, whether it be learning, understanding each other, or completing a task. This type of collaboration was described as making decisions together as part of the *bargain* of teaching and learning. Johan, a graduate professor, commented,

> If [students] didn't need help learning, they wouldn't need a teacher, right? If you already knew how to do something or if you could just teach yourself how to do it, then you probably would just do that. You wouldn't be sort of asking . . . You wouldn't be setting yourself up in this environment where you have a person who's trying to sort of help you figure it out. So I think part of it, making a connection to me, in that sense, that's just part of what I think of as the bargain of being teachers and students. The bargain is, I have the knowledge of something and you would like to learn about that or somebody else feels that you should learn about this, and so let's try to work together to help you learn about it.

In his description of the "bargain of being teachers and students," Johan gave students an active role in the interaction. He portrayed students as actively deciding whether or not to participate in their learning, thus choosing whether or not they would collaborate with the teacher in this process. Johan did not see teaching as a unidirectional flow from teacher to student; instead, he saw it as necessarily having two active agents—the teacher *and* the student—who together share responsibility for the outcomes of their interaction.

Similarly, Sidney, a high school teacher, told us,

> I learned to own who I was, which I did within the first year of teaching, and just say, "It's not like I'm a superwoman or a bad woman. I'm just a white woman from [New England] and this is what I know and what

I'm bringing to the table, and I'm honoring what you know and what you bring to the table. And so, let me teach you what I know that will help you gain access to power and to, you know, freedom, in a way because education is freedom. And you teach me what you know and how . . . and the ways in which you know it, so that I can be a more inclusive person in our classroom."

Sidney's comment illustrates a view of teaching that is equal parts teacher contribution and student contribution. It is through sharing their individual sources of knowledge that Sidney saw herself and her students achieving the ultimate benefits of education: power and freedom.

Evident in the examples of both Johan and Sidney is the thinking that the collaboration between teacher and student incorporates each of them as independent systems with individual contexts. These two systems form a unit—to use Johan's words, they make a bargain—and collaborate to create a teacher-learner system with a shared purpose of developing new knowledge.

This was expressed clearly by Marian, a teacher of a parent-toddler class who greets her students each year by saying, "Here we all are. We have different parenting styles. We have different ways, and yet we're all gonna work together, support each other . . . we're all on the same team here. We're here for the growing and the learning and the loving of all these children that are in the class together, and of each other."

Mutual Effects

This third type of AoI refers to instances in which teachers describe the reciprocal effects experienced throughout the teaching and learning process. For example, teachers share stories of times when they changed an approach, responded to, or were affected by the behaviors or feedback of students; in turn, their response caused students to shift their behavior or approach to learning. Delia's teaching process highlights this awareness. As she described it, she was "always out there looking for new strategies, techniques, or ideas to implement within the classroom to better [her] teaching, as well as to enhance the students' learning experience." During her teaching process she was "constantly looking to see how the students are engaging with the material and looking to decide whether or not [she] need[s] to shift the strategy that [she is] using or present in a different way, based on how they're reacting or interacting with the material, and maybe [she] need[s] to micro-unit it more or add a visual for different kinds of learners." Delia, a middle school teacher, explained that her teaching decisions were dependent on students' intentional and unintentional feedback,[2] and their continued learning was based on her responses to this feedback.

The reciprocal effect was described by Susan as a "dance." Implied in this metaphor is Susan's understanding that her pedagogical moves—the decisions she made in her teaching—were dependent on her student's moves and that, likewise, her students' moves are dependent on her own. She admits that as a graduate professor she is always learning: "Never ever ever ever do you, as a teacher, want to stop learning. It influences the newness of your approach, which influences the receptivity of your students and makes them as excited as you are." In understanding the mutual effects of the teacher-learner interaction, teachers seem to be aware that the teacher and learner are experiencing changes in their understanding, knowledge, and identity as a result of their coming together to create a new system.

Synergy

Out of the data emerged an unexpected pattern, one that shaped a fourth type of AoI I came to call *synergy*. In these instances teachers often described the vibe or energy created during a deep human interaction of teaching and learning. It was as though some teachers saw the various forms of interaction as leading to an experience of flow between students and teachers. After sharing that she found teaching rewarding, Liz, a middle school teacher, provided a series of examples to illustrate why she found teaching so rewarding, such as collaborating with teachers, solving puzzles with kids, and even developing curricula. But she finally concluded that it was

> just that feeling, you know, of things clicking, things just are clicking, things are clicking, you don't want to stop. You don't want [to stop], and the kids are like, "I don't wanna go to gym," you know, "I don't wanna go home," and you're feeling the same thing. And a lot of that is really about human connection. Are we connecting? . . . Are we just jibing? And it's the same with teachers. Am I sharp? Are we? And you know what, so much of it is when you get in the zone. There's no here, no be, time kind of stands still, you know, when it's that zone of creative work that you're doing.

This description of joined flow was not unique to Liz; similar descriptions from other teachers were peppered with references to the different awarenesses of interaction.

Peggy, a pre-K teacher, shared that it was important to find synergy with her students so that she could make decisions together with them. In her own words:

> I feel like that's a huge responsibility that these kids bring who they are, so it's my job to find that synergy, and that means I have to spend a lot of time getting to know each one, one-on-one, and then when I put them in a group, when I put them in a pod of four, that they're actually actively listening to each other. It is so powerful when they take on the role of peer teaching

with each other, or they learn how to actively listen to each other, and the synergy that I'm thinking about is in those group meetings, where you're talking about balance and they just start clicking off. I set up a little provocation or I say some provoking things, and they just start to click.

In this example, Peggy illustrated not only the synergy between teacher and student but also how this synergy can then influence the synergy among students.

First, [students] want to sort of please me and they want to be part of this thing with this teacher, right? And then all of sudden, they start connecting with each other. So that's when I feel the synergy, when the hands start to go up and then they start to bring them back to each other, keep bringing them back to each other, for the concept.

This insight is exciting. It opens new doors, hinting at a potential expansion of the teaching brain system that identifies the multiple interactions involved in teaching.

Awareness of Self as a Teacher

While learning may occur without teaching taking place (and often does, in all manner of settings and situations), teaching never occurs without the dyad of teacher and learner. As the leader in this interaction, the teacher gathers information both from the learner *and from herself* and processes what is necessary for the interaction to be beneficial for the learner. This integration of a teacher's own personal and professional contexts is often overlooked or absent in current practice and in contemporary educational reform efforts.[3] Yet somehow we readily see the importance of being self-aware in many other areas of our lives. For instance, in her book *How Toddlers Thrive*, child psychologist Tovah Klein shows how good parenting occurs when the parent is able to understand where the child is developmentally.[4] When a parent shifts his or her point of view and understands the world from the young child's perspective, the parent is better able to direct that child, calm her down, and ultimately help her self-regulate and cope with the world on her own. If the parent simply takes the adult perspective and tries to control the child or force compliance, the child will feel shame, have difficulty self-regulating, and never really learn how he feels and thinks. This kind of parent awareness is all about the parent being both self-aware and child-aware.

Knowledge of self is required in order to effectively understand the inputs and outputs that together create a comprehensive teaching process. While teaching is an interaction between teacher and learner, the act of teaching itself is filtered through the teacher's lens. Therefore the teacher's awareness of self as a teacher—in other words, a teacher's understanding of her personal context and how that context interacts with the student—is just as important to the teaching brain as the teacher's awareness of the learner.

Personal Context

A teacher's personal context is influenced by both internal and external factors. The internal influences on personal context can be described as traits (both phenotypic and personality) that are specific to a teacher and are, for the most part, static. Arguably very few traits are forever static, but for our purposes here you might say that these traits are "just who the person is." Phenotypic traits are those that are obvious and observable, such as gender, skin color, and height, while personality traits are characteristics such as being garrulous, being quiet and reserved, or being adventurous. For example, I have the phenotypic traits of being female; having brown hair, skin, and eyes; and being just over five feet tall. While I suppose I could change several of those traits, I consider them to just be who I am. I might describe my personality as humorous, passionate, youthful, and fun-loving.

Each type of trait has an important influence on the teacher's practice. When teachers become aware of how aspects of their self affect their view of students, they are in a better position to process the multiple variables that contribute to a successful teacher-learner interaction.

Similarly, personality traits offer important insights into how a teacher goes about teaching. And, of course, a teacher's emotional or affective state always exerts an internal influence on his or her awareness of self. Are you naturally high-strung, or are you laid-back? Do you get anxious at certain times of the day or depressed during certain times of the year? Teachers are no more immune to human emotional experiences than the next person. The attuned teacher, however, becomes aware of these emotions and either uses them or sets them apart from his teaching at any given point during an interaction.

The external factors affecting a teacher's personal context involve relationships with other people, culture or lifestyle, places, and objects or environment. A teacher's context is constantly informed by the world surrounding her. Among the relationships that can affect a teacher's personal context include those with family, friends, other teachers, and administration.

Think of your most successful teaching moment, an interaction in which your learner really got whatever the objective may have been. Perhaps this moment was in a school, or maybe it was an exchange with your child, a sibling, or a friend. What did you do? Why did you teach that way? What may not be immediately obvious is that it's quite likely you were teaching in a fashion that was somehow responding to the way someone else—a school-teacher, a cousin, a neighbor—once taught you. Whether the experience was positive, negative, unremarkable, or unforgettable, and whether you attempted to mimic it, avoid it, or adapt it, to varying degrees it shaped how you approached teaching in that moment of glory. In that teaching effort you then received either positive or negative feedback that caused you to inhabit a specific, unique belief system, and voilà—that's (in part) why you teach the way you do.

Even the weather or current events can influence a teacher's personal context and, in turn, alter the teacher's processing of sensory information. For example, a major flood, a war, or just a week of steady rain may change the mood of the teacher and how he or she perceives the students and herself during that period. Similarly, a teacher's personal context is a product of his ambitions, goals, and income.[5]

When a teacher processes information from his personal context, he is considering his own process of thinking (metacognition), how a student's response will make him feel (metaemotion), and how much information he can recall to help support a student (meta-memory). A teacher can develop a model of his own cognition, emotion, and memory—essentially a comprehensive mental model of the self as a teacher—making it possible to manipulate the information, store it, and recall it rapidly when needed to effectively respond to his students.[6] His model exists at the "meta" level, creating metacognition, metamemories, and meta-emotions; these are essentially comprehensive mental models of the self as a teacher.

Meta-processing

Let's look a little closer at the formation of this model of the self as a teacher. Similar to what occurs when the teacher forms a theory of the learner's brain, the awareness of self as a teacher requires a consideration of cognition, memory, and emotion. For example, a teacher's meta-memory enables him to know what knowledge he would remember if he focused on recalling it. When a student is struggling to figure something out and the way a teacher has explained it has not worked, meta-memory is what a teacher calls upon to try to devise another strategy that may work in this particular situation.

A teacher's meta-emotions provide the teacher with an understanding of her emotions about her student. Parents do this all the time. For example, when your child lashes out at you because he doesn't understand how to do his homework, do you react with a sense of frustration? Do you feel like your child is angry with you? With the teacher? When you can engage in meta-emotional thinking, you take this one step further and consider how all of that makes you feel, and therefore why you responded the way that you did. Meta-emotional thinking is a recognition that emotions cause certain responses. This meta-process is critical because it enables the expert teacher to manage the overwhelmingly complex and constant stream of sensory information taken in from both the student and teacher contexts.

Teachers' multiple meta-processes are dependent on their awareness of the learner. Expert teachers not only understand the cause of a student's behavior (theory of mind) but they also consider what their student knows (theory of memory) and what the student is capable of doing (theory of cognition). Expert teachers utilize meta-cognition—being aware of their level of awareness—to understand how they are thinking about what their students are thinking; these processes are nested within each other.[7] An expert teacher is engaging in meta-emotional thinking when she is aware of how she is feeling about her own teaching actions *and* how she feels about her student's feedback to those actions.

Awareness of Teaching Practice

Awareness of teaching practice (AoTP) is more common in our current paradigm of teaching. Teachers' awareness of their practice occurs during moments where they are cognizant of all the "stuff" that makes up their teaching—not just their pedagogy.

This could have to do with more tangible procedures such as routines, organization, and time management. However, it could also involve less tangible methods that they utilize to help them create their classroom culture and ensure continued learning.

Some of the strongest literature on teaching practice has come out of the Center for Advanced Study of Teaching and Learning, founded by Robert Pianta. Pianta and colleagues developed a tool for classroom observation called the Classroom Assessment Scoring System (CLASS).[8] The tool is meant to be used by a trained CLASS observer during a classroom visit in which the observer evaluates the emotional support, classroom organization, and instructional support available during interactions between teachers and students. This tool can be quite useful in thinking about teaching practice, both inside and outside the classroom.

As mentioned earlier in the discussion of awareness of learner, recognizing a child's emotional functioning is critical to being able to support the child's learning. While having a sense of this for an individual learner is critical to forming an appropriate AoL, it is also important to consider the climate of the class as a whole in terms of both the positive and negative factors contributed by all involved in the teacher-learner interaction. Sometimes this is as simple as thinking about how often you smile and laugh during your interaction with learners. Considering the pitfalls that a learner might encounter during instruction and planning for them ahead of time is also a practice that enables teachers to show learners that they care. A significant aspect of a teacher's practice is how flexible they are when unanticipated challenges arise. This is often a strong measure of one's practice. Are you willing to be flexible and even let the learner lead you and other learners, or do you consider yourself the sole teacher? Because teaching is an interaction with mutual effects, its practice should reflect an awareness of this. All teachers are also learners.

The organization of our teaching practice is more often

what teachers think of when assessing classroom practice. Understanding your approach in its entirety helps teachers reach the expectations they have set for learners and themselves. These expectations may seem specific to the classroom, but in actuality we set learning expectations all the time. Teaching is an intentional task; we teach for others to learn. Clearly acknowledging expectations helps a teacher plan thoughtfully, set routines, manage time, and be thoughtful about the best format in which to teach a topic, idea, or skill.

Teachers' practice is also closely connected to how they respond to students' queries, conversation, and shared information. Much of this was discussed in the section on the awareness of interaction, but here it's crucial to highlight a specific type of teaching response that is strongly linked to skill: scaffolding. It can't be emphasized enough how important scaffolding is to teaching practice. Offering students support throughout their entire learning engagement is critical. Sometimes this just means giving them a hint; other times it means we offer an answer and ask them how we got it. It can even mean just letting them work through it on their own for a while so they can have their "aha" moment. There are times when we need to keep students going by encouraging them with many accolades, and other times where we need to draw a hard line and tell them they need to try harder. Whatever the case may be, scaffolding is an essential component to teaching practice. Becoming more aware of how to best manage it for ourselves and our learners helps improve our teaching.

Awareness of Context

As with the awareness of teaching practice, many aspects of awareness of context (AoC) intersect with elements defining other awarenesses. All teachers and learners exist within their own contexts, and therefore if a teacher has a strong awareness of himself as a teacher (AoST) and a strong awareness of learner

(AoL) he is more likely also to be aware of his own personal context and that of the student. Yet context does require its own distinct category, because in order to delineate that context a teacher must reach beyond the individuals involved in the direct teaching interaction and consider how the larger context—a macro view, so to speak—affects the teacher-learner interaction. Many things can factor into this awareness. What follows are a few sample considerations.

Let's consider external influences that exist entirely outside of the immediate context of the learner and teacher. In a school setting, these influences would include the principal or director, other teachers, and even the school philosophy. All these external factors impact the teaching interaction, whether directly or indirectly. Principals likely set the rules of the school, other teachers impact the students' emotional and academic growth, and the school philosophy can dictate how a child dresses and engages with friends and peers.

As we witnessed with the No Child Left Behind Act, even the federal government has a great impact on the teacher-learner interaction. The latest developments have come with the Common Core State Standards, which, although not a product of the federal government, are closely tied to federal education policy and dollars. Local, state, and national governments, in their efforts to close the achievement gap and fulfill the right of every American child to a quality education, have often created standards that short-circuit a full understanding of teaching. These policy initiatives have also come to affect how parents teach children at home. More and more parents are convinced that the goal of teaching their child is to help the child score well on standardized exams in school.

Stepping outside the tense politics of education into the equally tense realm of current events, what is happening socially, politically, and economically in the larger world has great bearing on our teaching. No doubt in the wake of so many school shootings in recent years parents across the country have spent a little more time teaching their child about safety, perhaps even about violence and gun control.

It makes perfect sense that our larger context affects how and what we teach. Teaching is one of the primary ways humans combine knowledge, thus expanding our collective knowledge and coming to better understand one another.

Why Awareness Matters

As we've seen in earlier chapters, teachers who practice student-centered teaching create mental models of a particular learner. The expert teacher typically has a more nuanced and complete model of a student's disposition, learning style, emotional experience, and other facets of the student's learning context. This awareness of learner includes how the student might respond to various teaching choices (theory of mind) as well as theories of the child's capabilities (theory of cognition), how the child's learning might be affected by his or her feelings (theory of emotion), and even how far the teacher can push the child to recall information that the teacher believes is stored in the child's memory (theory of memory). Teachers use all of this varying information to command their "efferent limb"—to continue the metaphor of the nervous system—and enact a student-centered response.

A teacher's responses to a student are limitless in that they are always a combination of what any individual teacher brings from her personal context as well as her multiple meta-processings of self. This is well illustrated by classroom teachers who are empowered to create their own lesson plans or classroom seating, or by the many studies showing that teachers tend to think boys are better at math and science than girls. However an awareness of self as a teacher (AoST) encompasses much more than single

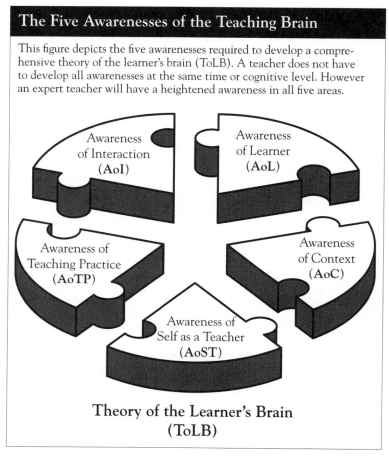

The Five Awarenesses of the Teaching Brain

This figure depicts the five awarenesses required to develop a comprehensive theory of the learner's brain (ToLB). A teacher does not have to develop all awarenesses at the same time or cognitive level. However an expert teacher will have a heightened awareness in all five areas.

Awareness of Interaction (AoI)

Awareness of Learner (AoL)

Awareness of Teaching Practice (AoTP)

Awareness of Context (AoC)

Awareness of Self as a Teacher (AoST)

Theory of the Learner's Brain (ToLB)

Figure 5

observable behaviors. A teacher who is highly self-aware also recognizes that her lens completely shapes how she views her students and therefore her students' learning brains. It's hugely important never to forget that the theories we form of students are in fact *theories*, not truths. Often we act upon them as if they are facts, leading to a dangerous rigidity in our understanding of ourselves and our students.

Student-centered teaching without a sense of self-awareness

as a teacher is an ignorant act, as it sets up the teacher as truth maker. We may think of this as altruism, but it is the opposite. If a teacher does not acknowledge his or her self, the teacher almost automatically treats theories about their learners as fact— and leads the public to wonder why well-intentioned teachers still don't have successful teacher-learner interactions. We might resort to blame, either of the teacher ("She's not following best practices!") or of the student ("He's unmotivated!"). In reality the problem is much more complex.

A teacher who utilizes multiple awarenesses (of self, learner, interaction, teaching practice, and context) to form a theory of the learner's brain (ToLB) will likely produce a more comprehensive, accurate theory. Why? Because the teacher will recognize that when the teacher-learner interaction fails, *it's because the teacher's theory was incorrect*. As a result of such deliberate practice, expert teachers look to evaluate their awarenesses to identify where exactly they went wrong.

- **Was the teacher's theory formed on an incorrect awareness of the learner?** For instance, the teacher may have thought the learner was shy, when in fact the student is confident and extroverted, but in the student's culture it is disrespectful to ever look an adult in the eye or speak proactively.

- **Did the teacher perceive the student as apathetic because of a weak awareness of self?** Let's say a teacher's culture contains norms of being passionate, forthcoming, and talkative—that is, in this culture motivation is exhibited through external behaviors. Perhaps these norms are culturally or personally abnormal for the student.

- **Is it possible that the teacher's awareness of the interaction was different from the student's?** The teacher thought a level of synchrony was reached, but it was an out-of-tune personal flow that ignored a student's unintentional feedback.

- **Was the teacher aware of the school context?**
  This school's philosophy may include the very strong belief that student learning is best assessed via standardized exams.

Such insights allow the teacher to home in on the breakdown and revise his or her theory. An expert teacher considers each failed interaction as an occasion to learn more about a possible solution—as an opportunity for deep reflection and development. Without acknowledging in a meaningful, systematic, and complex way the impact that the teacher has on the teaching interaction, we are unable to see clearly the origin and development of a particular teaching response and what changes could or should be made to the teacher's processing.

If teaching is an interaction, having an effective impact on the teaching response requires that all parts of the interaction be taken into account. Teaching is not a linear process of inputting knowledge into the learner. Teaching consists of at least two variables, the learner and the teacher, and each of these variables is in turn defined by a practically infinite number of variables. Teachers who are effective respond both to their sense of self as a teacher and to their students. Their teaching responses are likely to be more successful because they are able to adjust to the "real" learner rather than to a generic, baseless, or simply incorrect theory of their learner.

This kind of processing and response to a learner's needs stands in direct contrast to pedagogy that is based solely on best practices and directive behavioral techniques.[9] It also differs from certain progressive, student-centered instructional frameworks that suggest responding solely to the learner.[10] Though it is of course helpful to consider the learner, focusing solely on the learner precludes a complete understanding of what happens when teachers process. It means that some students and teachers always get lost in the cracks, which in turn leaves the door open to repackaged quick fixes touted as innovative education reform. Again, rather than focusing so much attention on the teaching response alone, the more deeply we understand a teacher's processing, the greater the benefit to everyone who is a part of the interaction. Going back one step further is not just helpful or good but necessary; it is in the processing that teachers make decisions about how to respond. If we hope to positively affect the responses of teachers, the key is in adjusting the processing. A response-only approach enters the interaction too late to create any real change.

THE POWER OF DELIBERATE PRACTICE

The necessary awareness of process emerges from extensive, deliberate practice. As Sanjoy Mahajan, a writer for Steven D. Levitt and Stephen J. Dubner's Freakonomics.com points out, "Deliberate practice requires sustained concentration, and the rewards are subtle and apparent only in the long term. Thus, one needs motivation in order to enter into and sustain the hard work of deliberate practice. But the learning happens not simply through putting in the hours, but through doing so intelligently."[11]

In the most basic way, expert teachers are deliberate about how they reflect on their practice. They are aware of the multiple areas that they must consider in order to hone their craft: the learner, the interaction, the context, their teaching practice, and their overall lens as a teacher. Expert teachers ask themselves: "Why did I become a teacher? What are my assumptions about teaching? How do I approach teaching? What do I expect to get out of this enterprise?"

I imagine that most classroom and professional teachers do not consciously deliberate on questions of "self as a teacher" on a daily, weekly, or even annual basis—perhaps because of a lack of will, perhaps because of a lack of time. Certainly many of us ruminate about these questions when we first imagine teaching, when applying for a job, or when considering graduate school. But such questions are often pushed to the side or dropped entirely once we've become teachers, often arising again only in moments of struggle or success.

However, expert teachers must be just as deeply aware of the self as they are of their learner, because teaching is unlike learning in that it cannot be an independent endeavor. When we suggest that we have taught ourselves something from a book or online, we are actually just stating that we have learned. Teaching, on the other hand, is an interaction between two people, the teacher and the learner, and the information that is inputted and then processed by the teacher to form a teaching response cannot be solely self-generated. The higher the level of cognitive thought that goes into the processing, the more the response can enhance the teacher-learner interaction. It's quite intuitive and works like any number of interactions: the more thoughtful you are about your response, the more likely it is to be effective. The classroom teacher who is an expert earns that distinction because of how she has thoughtfully chosen to develop her teaching awarenesses.

SPEED-DATING A BOOK: THE RESPONSES OF AN EXPERT TEACHING BRAIN

An expert teaching brain is one that has maximized its brain processing to form heightened teaching awarenesses (AoL, AoST, AoI, AoTP, AoC).

Imagine a classroom that is embarking on a standard student-centered lesson of choosing a book that is a "good fit." Typically this lesson is taught such that it supports Gardner's notion of multiple intelligences, with a variety of approaches to how a student can choose an appropriate book.[12] Teachers are trained to present "mini-lessons" that last only ten to fifteen minutes in order to hold student attention. They use large white chart paper to brainstorm ideas so that students feel they have an investment in what will become the class's protocol for choosing a book. Class charts are limited to three colors or less to avoid overstimulation and confusion for students. All charts must include borders so that students can focus on the important information at the center of their visual field. These large student-centered charts are then displayed in the room so that students can enjoy the print-rich environment meant to support and stimulate their language acquisition.[13]

Students are also instructed to look at the picture on the book cover (for visual stimulation), read the back cover (to gauge their interest level), and use the five-finger rule (if there are five words on a page that you do not know, then it is too challenging a read). Every year, the teacher carries out the same lesson, which is packed with "best practices," that is, practices that have been identified as "successful" because "successful" teachers utilize them.[14]

Now reimagine this student-centered experience through the lens of an expert teacher, Claire, with a fully formed ToLB. In this classroom, when the teacher asked her students, "How do you choose a good book?" one young boy, Jackson, replied, "I date the book."

Claire, an eight-year veteran and literature major, chose to take advantage of this unique and clever suggestion. She revealed that in college she would often look up the date a book was written and consider the current events of the time, the biography of the author, and the time period it was representing. This would reveal a lot of information about theme and plot of the text, which in turn could help her decide whether it was a good fit for her (AoST).

On the basis of her experience, Claire perceived the boy's

suggestion as quite sophisticated, so she encouraged him to say more about his suggestion. However, Jackson corrected her quickly: "No, I *date* it. You know—I take her out on the town, bring her to dinner, and see if I'd rather go watch a movie or just go back home and stay up talking with her all night."

The difference between a student-centered response and a teaching brain response is illustrated by the sequence of events that followed. Rather than just having Jackson share his example and then moving forward with the evidence-based lesson, Claire replied (with a hint of her typical sarcasm [AoST]), "Well, aren't you the ladies' man, Jackson! Why don't you tell us how you know she's the one?" Jackson continued to share his method for choosing a good-fit book. Meanwhile, the sensory information he provides constantly interacted with Claire's personal context; rather than simply laughing along with the class, she processed the experience and allowed it to drive the creation of an entirely new and unexplored interaction. The result of this dynamic interaction was a new, collaborative class project called Speed-Date a Book, in which students chose two of their six summer readings and had to figure out how to get their peers to agree to "date" their book (AoTP).

The project represented an expert teaching brain response because it was a product of all her teaching awarenesses. This project incorporated the student context because it was student generated (AoL) and connected to a contemporary concept that was highlighted in a recent adolescent blockbuster movie (AoC). However, it still constituted a traditional student-centered lesson. Students determined their books' best attributes (theme, plot, characters, etc.) and planned how to represent them attractively to others. Students then used the vehicle of speed dating: they sat across from a semi-interested peer and had two minutes to convince the peer to date their book (AoTP). A successful book date resulted in a student adding that book to his or her reading list.

This project tapped into the students' theories of mind, cognition, and emotion, giving the lesson a critical "hook" or "trigger" to help increase the students' motivation and understanding.[15] It also incorporated the teacher's context, her preference for literature, the careful use of humor, her awareness of current trends (AoST), and her ability to rapidly assimilate the student context to her own in order to create synchronous teacher-learner interactions (AoI). Moreover, the project met all the same content and skill requirements of the previous best-practice models mandated by the school (AoC), but in the reimagined scenario the educational experience was a dynamic interaction between the teacher and students, grounded in a ToLB. Student needs were still being met as the teacher experimented with a new project, even though this dynamic process would clearly controvert current trends in education for standardizing classroom interactions to enhance reliability, accountability, and transparency.[16]

Furthermore, the freedom to invent unscripted teacher-student interactions motivated the teacher: it engaged her interests,

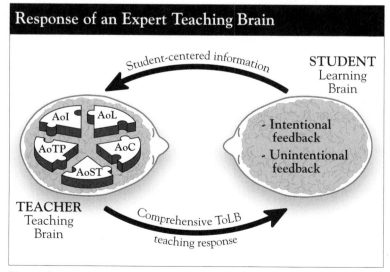

Figure 6

challenged her ability to plan a new project, and extended the bonds that she shared with the class. They all enjoyed a warm-hearted laugh over Jackson's unique way of choosing a book. The resulting project transcended the curricular requirements of providing evidence of students' summer reading for the purpose of assessing their reading level. Claire used student responses not only to assess her students' reading level but also to evaluate the success of her project design, creativity, and teaching skill. These are important validators and motivators for continued efforts over the long term.[17]

This type of complex teaching brain response produces many benefits for student and teacher. It solidifies the teacher-student bond, encourages the class to work as a unit, provides both students and teacher with feedback, and moves them toward synchrony of action and thought.

TEACHING TOWARD SYNCHRONY

Inevitably a teacher's response is an effort to join with her learner. Humans are social beings, and this is quite apparent when it comes to teaching. Teaching is an effort to become one as a society. It enables us to have common thoughts and also to create new knowledge together. This is why civilizations since the beginning of time have looked first to education when building their social system. All of the great empires—Greece, Rome, Egypt—developed an education system that served as the bedrock of their civilization. Teaching connects people and helps to develop, shape, and define a society.[18]

Let's return to the nervous system metaphor of sensing, processing, and responding. A teacher's ability to create a theory of the learning brain is distinct from simply having multiple awarenesses. Many people are able to form awarenesses; it's not just expert classroom teachers or even just adults who are capable of this. What distinguishes the expert teacher from the rest is the ability to utilize those awarenesses to form a ToLB, and the effectiveness of the teacher's response built on this theory.

How a teacher responds reflects a degree of synchrony of action and thought between teacher and student. For the teacher-learner interaction to work well, there needs to be a feedback loop between teacher and student, and the feedback needs to indicate success. When this occurs, we reach a level of synchrony in which teacher and learner have joined in knowing. Then the process begins again, renewing itself.[19]

Furthermore, this cycle of recursive processing is a continuous loop in which the student's learning brain responds to the teacher's actions and sensory input. This provides critical feedback that the teaching brain processes to adjust the teacher's responses. This constant feedback loop is, I suggest, the source of the intangible synchrony that occurs between teacher and student. Kazuo Yano and Alexis Kent call this the give-and-take between teacher and student.[20] The synchrony it produces is hard to define, but you know it when you see or feel it. It is the flow that drives creativity and higher human thought.[21] It is the X factor that makes in-person teacher-student relationships irreplaceable—because this feedback loop is based on the full body of interaction and not simply voice, visual, or textual interactions alone. The teacher-learner interaction is the engine behind the synchronous educational experience that characterizes the best teaching and learning brains.

For teachers, the dimensions of the teacher-learner interaction are not black-and-white, nor are they fixed. Teachers are human and therefore are continually changing, growing, and adapting, with their own fits and starts. The process of becoming an expert teacher ultimately begins in a place of awareness—knowing that these dimensions exist and that they inevitably impact the process of teaching.

7

DEVELOPING THE TEACHING BRAIN

Chapter 3 introduced a skill scale that depicts how learning, as a skill, becomes more complex over time. Younger children move from very concrete reasoning to more abstract thinking as their skill develops. This theory, known as dynamic skill theory, also posits that with support and scaffolding—usually by a teacher or parent—learners will reach their optimal level of understanding.

USING THE SKILL COMPLEXITY SCALE TO UNDERSTAND TEACHING

Similarly, the skill levels of teachers vary depending, in part, on the presence or absence of support. In his work, Vygotsky defined the zone of proximal development (ZPD) as the difference between what one can do with support versus what one can accomplish independently. Teachers also have zones of proximal development, underscoring the dynamic, context-dependent, varying nature of their teaching skills over time. The dynamic skill scale is a helpful tool for measuring cognitive complexity. It allows us to understand the dynamic and complex develop-

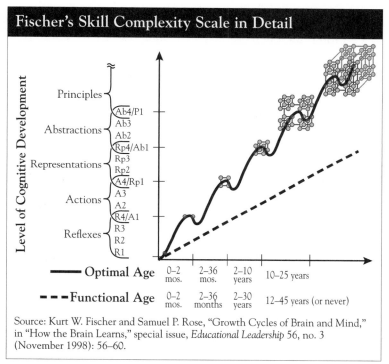

Fischer's Skill Complexity Scale in Detail

Level of Cognitive Development

Principles

Ab4/P1
Ab3
Abstractions
Ab2
Rp4/Ab1
Rp3
Representations
Rp2
A4/Rp1
A3
Actions
A2
R4/A1
R3
Reflexes
R2
R1

—— **Optimal Age**   0–2 mos.   2–36 mos.   2–10 years   10–25 years

– – –**Functional Age**   0–2 mos.   2–36 months   2–30 years   12–45 years (or never)

Source: Kurt W. Fischer and Samuel P. Rose, "Growth Cycles of Brain and Mind," in "How the Brain Learns," special issue, *Educational Leadership* 56, no. 3 (November 1998): 56–60.

Figure 7

ment of teaching as a skill of all humans by bringing the ZPD for teachers alive and making it tangible.

Fischer's skill scale shows how learning, in the presence of scaffolding, enables the learner to reach beyond his otherwise functional range of understanding to a higher, more optimal level of understanding. The skill of teaching exhibited by any of us also goes through a process of development responsive to the presence or absence of support. Though professional teachers may be given direct support through mentoring, professional development programs, or curriculum aids, the most prevalent support is the learner.

The feedback that learners give is most often what we use to know whether we are doing a good job. Building our awarenesses will help us to understand this feedback more directly so that we can intentionally modify our teaching toward achieving synchrony with our learner. This is not an easy task. As we develop, our trajectory will likely present fits and starts, spurts and plateaus, where we strengthen one aspect of our teaching but struggle a bit when we begin to develop a new one.

Fischer's scale offers a framework for understanding where one's teaching level is in terms of one's *own* growth and development—not as a method of assessment or comparison against other teachers. A fundamental notion is that individuals construct skills in order to participate in specific tasks within their context.[1] Consider again teaching as an evolved skill. Perhaps the original instinct was to teach for survival, but today parents teach for love, for connection, and to help their child belong. Teaching is an adaptive skill that we have continued to develop over thousands of years.[2] Like any skill, it was constructed to facilitate growth and advancement.

While the last three tiers of Fischer's scale are best suited to measuring learning observed in students, the situation of teaching is different. Teaching requires more sophisticated skills than learning because it is always an interaction. The teacher has to be cognizant of the learner's mind; therefore she must already have reached the tier of representations. At the very least, a teacher must be able to form a theory of mind in order to teach a learner. Teachers have to develop through all four levels of increasing complexity within a tier, mastering the associated skills, before they can move on to another tier.

Someone at the first level of any tier on the skill scale exhibits the knowledge that is characteristic of that tier (reflexes = R1, actions = A1, representations = Rp1 or abstractions = Ab1, principles = P1). However, these are the simplest expressions of that knowledge. The second level of any tier requires that a person be able to organize this simple display of knowledge into mappings. For example, the undergraduate student taking a humanities course suddenly sees the interdisciplinary thread among science, history, and literature during the twenty-first century as Western world economies shift from industry to technology. When the student can coordinate multiple mappings to create a complex system, he or she has achieved the third level of understanding within that tier. The ability to conceptualize a system of systems, in which systems within the tier can be coordinated and related, indicates that the fourth level has been reached, and the learner will have graduated to another tier. Therefore the first level of any tier is the fourth level of the preceding tier (A4 =Rp1, Rp4 = Ab1, and so on). In the third and fourth tiers, thinkers not only are able to understand the content of the changes taking place during the twenty-first century, for example, but can pull back and see the larger contextual framework, as one might expect from Kuhn's theory of paradigm shifting.[3]

The movement through each tier is especially difficult in teaching because the object of a teacher's learning is primarily a learner. Teachers improve their skill based on how well they understand or are aware of their learner (AoL), but learners and their learning are forever changing. For teachers to reach a point at which they can form a theory of an individual learner's brain (remember that a ToLB can be formed only with the help of multiple awarenesses), they must be functioning at the final level of abstractions (Ab4 = P1). Creating a ToLB means that the teacher is able to form a system of systems and create a new principle for understanding the learner. Doing this with multiple learners indicates that the teacher has reached the P4 level and is able to relate systems of multiple principles to one another. This level is achievable only by expert teachers who form multiple ToLBs, one for each learner, and act upon them all seamlessly within one teaching session.

Other teachers stay within the lower levels of abstractions or representations. Teaching can still occur at the tier of representations, for example. This kind of teaching occurs when teachers have an understanding that a learner is a representation of a system of learning. The teacher is able to make guesses about what a learner knows and adjusts his or her practice to support the learner's learning. This is called contingent teaching (as discussed in chapter 1) and can be achieved as early as age seven. Teaching within the tier of representations often takes place without any awareness on the teacher's part of the entire system of teaching as an interaction between teacher and learner. The teacher teaches to the representation of the learner, regardless of whether it is the correct ToLB for the learner with whom they are working (something that can be seen in the example of Logan in chapter 5). If this representation of the learner is based on ideas about what is "typical" or the "norm" or "average," it is still possible for the teacher to have successful interactions, but it is much less likely.

For example, a teacher may have decided, based on some statistical norm she has heard or read about, that her white male student should be better at math than a girl of color. The teacher teaches with that representation in mind, regardless of the learner as an individual. If the student is learning Algebra II with ease, it may appear that the teacher is teaching well. However, the apparent goodness of fit may be more a result of the individual student's skills than of the teacher's understanding of the needs of that particular student. In other words, the teacher may not have formed a correct ToLB. As a result, a teacher could unconsciously or consciously arrive at misguided—even harmful—interpretations of a student's skill or lack of skill. If, for example, the teacher assumes that a student of color is not going to be skillful at math, she may unconsciously lower her expectations of that student and transmit the lowered expectations to the student in any number of ways; the student might then not be offered an opportunity to be challenged in her learning.

Recently Strauss and Ziv published a comprehensive matrix of the strategies that children use to teach, with supporting examples from various studies in the field.[4] (See Appendix C.) While we do develop more sophisticated skills over time, we don't abandon earlier teaching skills, such as demonstration and explanation. Instead, the multiple teaching awarenesses we build enhance our ability to use those skills effectively. Expert teachers will still utilize demonstration, for example, but the thinking behind *why* they have chosen that action is much more cognitively advanced than that of a toddler.

I'll illustrate this with various teaching scenarios based on specific developmental levels. In each scenario the teacher is utilizing a range of skills from the skill development scale. (I haven't included skills that are beyond the students' developmental capacity, such as age level, brain development, etc.) I'll point out the thinking involved in *why* the teacher is using that skill, to highlight the increase of sophistication from a toddler teaching a peer to an expert classroom teacher working with a group of students. This is the area that is most forgotten in the literature on teaching, but I want to draw attention to the processing—to what exactly is going on in the teaching brain when we are teaching.

The skill that requires the least amount of cognitive complexity is demonstration. Demonstration can be something as simple as pointing to an object that someone needs in order to fill a gap in the learning process. To have sensed that the learner had a knowledge gap in the first place, the teacher must have a ToM. The teacher then offers support by demonstrating something that could help the learner fill this knowledge gap. Put simply, the teacher has discovered that he or she knows something that the learner does not. For more explanation of this low-level teaching, refer back to the animal and infant teaching examples in chapter 1.

In the next example, let's focus on the first three levels of representation. We can see these most clearly in the preschool years and middle childhood.

TEACHING WITH REPRESENTATIONS IN MIND: TWO TO SEVEN YEARS OLD

At seven (Thomas), four (Jordan), and two (James), my nephews make the perfect guinea pigs for my research on teaching. Recently my whole family stayed together at a lake house for a week. I designed all sorts of fun activities for us to engage in: relay races, scavenger hunts, canoeing. Unfortunately it rained for the first three days. One rainy day was fine; we did craft projects. On day two, we cooked. But on day three, I was stir-crazy and not much in the mood for leading another activity. What I didn't expect was the sheer resilience of my nephews, who were determined to find something fun to do. The house we'd rented had the game Mouse Trap, which is labeled for children from six to twelve years, and which none of the boys knew how to play. Thomas was fascinated by this game, and even though his mother and I weren't in the mood to teach the kids how to play, Thomas was determined. He decided to teach his brother and cousin how to play.

Thomas opened the box and began pulling out all of the pieces. He convinced his four-year-old brother to lay the game board on the coffee table: "You know how to do it, Jordan. Just lay the board down like you always do on game night." (Here Thomas was exhibiting an AoL by relating the task to something Jordan had done in the past). Then Thomas proceeded to try building the mouse trap.

Watching three children seven and under trying to put together a Rube Goldberg machine is quite a sight. When Jordan couldn't figure out the right place to put a rubber band, Thomas said, "We just have to put all of the parts on the board according to the picture, see? Whoever gets the most parts in wins" (here he was demonstrating and explaining). Jordan kept fooling with the rubber band and eventually discovered he could put it around a handle, pull it, then release it so that it shot across the board. Thomas stepped in at this point and said, "No, not like that. Look, each piece has to go on top of its shape on the board. The color of the part is the same as the color on the board. Look at the box. It's like this." (Here Thomas was demonstrating and explaining, but without learner involvement.)

His instructions to James were quite similar. But at just two years old, James couldn't contribute much to this exchange. (This revealed Thomas's lack of an AoL for James.) James was happy just to name the parts. "Mouse," he kept saying as he grabbed the piece. Thomas kept telling him to pull all of the pieces out and put them on the board. "Look, James, see the colors. Blue goes with blue and red goes with red, like this." James of course did not quite understand, but Thomas just kept repeating the instructions, doing his best to include James (again demonstrating the lack of an AoL for James).

Thomas had never been taught how to teach, but there he was teaching. His teaching was a natural human response to the situation. He was faced with learners who knew less than he did, and only through teaching could he close that knowledge gap (revealing a basic ToM). This was of particular interest to Thomas because he wanted to play the game. At his developmental level he was able to demonstrate and explain. However, his process was limited to what he could achieve within the representations tier of the skill complexity scale (Thomas was at Rp3, Jordan was at Rp2, and James was at A4/Rp1).

Thomas did demonstrate and explain, but he was not able to modify his teaching based on a theory he had formed of Jordan's and James's learning. As a learner, Thomas may have reached the early level of systems representation (Rp3) and was able to understand how to put together the elaborate mouse trap, but he was not at Rp3 as a teacher. A teacher's ability to climb the complexity scale is also dependent on how aware he is of his learner and of the factors affecting that person's learning. In this regard Thomas was only beginning to form single representations (Rp1) of his learners. He had a very limited awareness of learner (AoL) for Jordan and did not have an awareness of himself as a teacher (AoST), his teaching practice (AoTP), the external context (AoC), or his interaction with Jordan (AoI). Unfortunately, this meant that he could not modify his teaching to more effectively support his learners, Jordan and James. Thomas just wanted to play the game! His unique teaching style was based on his own personal context and ambitions interacting with the contexts of Jordan and James. But he was not yet able to form abstractions that would have supported him in synchronizing all of their contexts.

Recognizing this example of Thomas as a motivated young teacher, with self-interest but with limited scope, is a revolutionary approach to understanding teaching across all humans. Thomas does not teach because he hopes to help his brother score better on an exam, nor because he thinks he's in service to his cousin. Instead he was practicing and developing his natural skill of teaching because he wanted to play a board game. In order to ensure that he could play the game, he had to teach. The game may not have ended the way it was supposed to and may not have been played according to the rules, but they did play and had quite an enjoyable time of it.

In order to get a better picture of what Thomas was thinking, let's take a look at the figure below. Sticking with the nervous system metaphor, we can see what Thomas was sensing, the type of processing he achieved, and therefore how he responded. (Again, the term "spinal cord processing" is not being used literally; instead it helps to explain the lower-level, simplistic processing involved in teaching before more complex awarenesses are formed within the abstractions tier of the skill scale.)

Children Teachers

This figure depicts teaching within the representations tier of Fischer's Skill Complexity Scale (Figure 7). Refer to Appendix C for an overview of proto-teaching through contingent teaching exhibited.

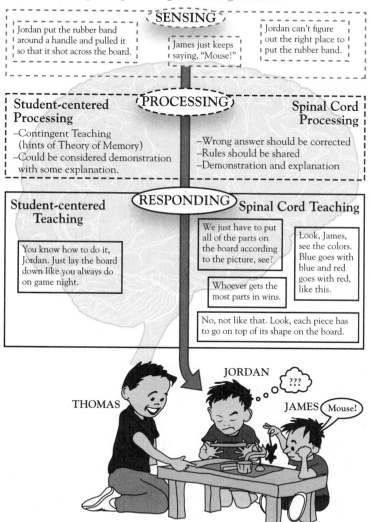

SENSING

Jordan put the rubber band around a handle and pulled it so that it shot across the board.

James just keeps saying, "Mouse!"

Jordan can't figure out the right place to put the rubber band.

PROCESSING

Student-centered Processing

–Contingent Teaching (hints of Theory of Memory)
–Could be considered demonstration with some explanation.

Spinal Cord Processing

–Wrong answer should be corrected
–Rules should be shared
–Demonstration and explanation

RESPONDING

Student-centered Teaching

You know how to do it, Jordan. Just lay the board down like you always do on game night.

Spinal Cord Teaching

We just have to put all of the parts on the board according to the picture, see?

Whoever gets the most parts in wins.

Look, James, see the colors. Blue goes with blue and red goes with red, like this.

No, not like that. Look, each piece has to go on top of its shape on the board.

THOMAS   JORDAN   ???

JAMES Mouse!

Figure 8

PARENT-CHILD TEACHING: TOWARD BUILDING ABSTRACTIONS AND IMPROVING AWARENESSES

Tracy is a mom of two daughters, Anne, nine years old, and Bea, six. Tracy had noticed differences between the girls early on (ToM), before they had begun their formal schooling. Tracy read aloud to both daughters, but Anne took to reading right away, as Tracy herself had done as a child, while it took Bea longer (AoL—theory of cognition). Tracy read aloud to both daughters, but Anne learned how to read much sooner than Bea; Anne began reading aloud to herself at about five and a half years old. Soon after she was writing, sitting down and doing her simple homework tasks, and in general reminding Tracy of herself: a "natural student" who loved to learn. But Tracy saw that Bea's approach to reading, writing, and learning was different (AoL— theory of cognition).

At first, Tracy thought Bea just wasn't paying attention, or perhaps that she was jealous of her older sister so she wouldn't try as hard (AoL—theory of emotion). She noticed that both girls loved hearing the stories and listening; they showed strong comprehension, followed story lines, and had insight into the characters. However, at around age six Bea still seemed to be struggling to decode and keep track of words as she worked her way through a sentence. It occurred to Tracy that how she interacted with Bea had to be different from the way she interacted with Anne (AoI). So Tracy shifted what she was doing with her younger daughter: She gave Bea support so that her reading automaticity—the automatic retrieval of a familiar or known word that allows for reading without having to struggle or think about it—and other early decoding skills became instinctive; she helped support her vocabulary by practicing more and more sight words with her, such as "also," "any," or boy" (AoTP). At around age seven and a half, Bea's reading took off, and soon she was reading as well as or better than most of her peers.

In this example, Tracy, an attuned and caring mom, had made

an assumption that her daughters were just alike as learners—and that they were like her (AoST). By paying attention to the differences between her daughters as learners and herself as a teacher and learner, however, she was able to provide the right support. She adjusted her interaction as a teacher and helped her younger daughter grow and flourish as a reader.

A few years later, when Bea was in sixth grade, she began to struggle with the advanced math curriculum. Bea felt strongly that her problems were due to the fact that her teacher "hated" her, and she told Tracy horror stories about the ridiculous demands that the math teacher was putting on Bea and how nothing she ever did was good enough (AoL—theory of emotion). Homework time became a real struggle: Bea didn't want to do the work, because she just didn't think it mattered. Tracy found herself sympathetic to Bea's plight. She remembered the damaging interactions she had with her own algebra teacher when she was in high school (AoST—personal context, past experiences).

Rather than feeling she and Bea were opposites, as had happened while Bea was learning to read, now Tracy actually felt they had had similar student experiences. However, Tracy didn't know how to help Bea out of this situation, and Bea's grades dropped, along with her motivation. Because much future progress depended on Bea understanding middle school math, Tracy knew she couldn't afford to let this slowly work itself out. After meeting with the math teacher, Tracy realized that Bea had a very important detail wrong: the teacher didn't hate Bea. In fact, she felt like a failure because she hadn't been able to figure out how to help Bea improve her math skills.

The problem was that the teacher didn't understand who Bea was as a learner. Just as Tracy had done while trying to support Bea's reading years earlier, the teacher was trying to teach Bea based on an incorrect ToLB for her.

Tracy's terrible memories of her algebra teacher quickly vanished, and she was able to recognize that she had some more work to do in figuring out why Bea was putting up this wall to the math teacher that prevented both of them from getting to know each other. She also knew she had to sort out how her own past was preventing her from forming a complete ToLB for Bea (AoI).

As we can see represented in the figure below, Tracy was utilizing cognitive teaching skills to help her form awarenesses both of herself and of Bea as a learner. Though at first her theory of emotion for Bea was incorrect, she later modified it as she became more aware of herself as a teacher and recognized that her past experiences had clouded her view of Bea's struggle. Bea's issues with math required that Tracy develop a more advanced AoST so that she could improve her ToLB for Bea.

One of the most challenging realities in education is that a teacher's development is based on his abilities as both a learner and a teacher. While a teacher may be capable of reaching abstract systems in his own learning, he must also be able to reach this level in forming a ToLB from his awareness of learner (AoL)—and generally not just for one student but for many students at once. Though the diagram does not show this, you could imagine each of the five awarenesses having its own skill development scale. As one awareness develops, it helps another to develop; only when they have *both* passed the fourth level within a tier, though, can they come together to form a correct and complete ToLB. That is not to say that the teacher cannot be successful with a partially formed ToLB for a student, or even with a ToLB that is incorrect (though in that case it's a risky game of chance). Teachers are able to be successful all the time with incorrect, standardized, or overly generalized ToLBs. It's often only when struggles arise and our practice proves ineffective that we realize our ToLB is wrong. When our awarenesses are not advanced enough to support us in shifting our ToLB, then we are often left with failure both in our teaching and in a student's learning.

Adult Parent Teacher

This figure depicts teaching within the abstractions tier of Fischer's Skill Complexity Scale (Figure 7). Refer to Appendix C for an overview of the conflicting mental models formed in this teaching example.

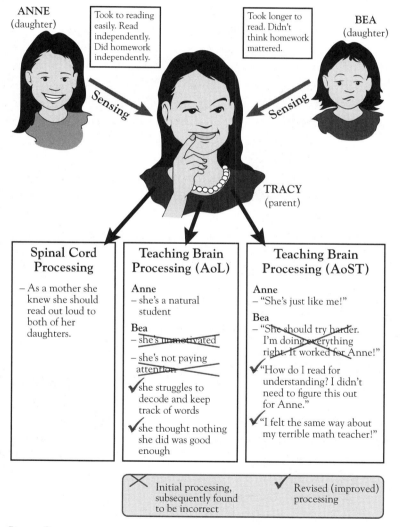

Figure 9

MIDDLE SCHOOL STUDENT AND TEACHER: ABSTRACT SYSTEMS AND COMPLEX PRINCIPLES

The original version of the skill complexity scale did not include the four levels for the principles tier (P1–P4). Researchers thought that the earliest one could reach that tier would be ages thirty to forty-five and that many would never reach it, so they did not do additional studies on that tier, instead believing that most learning ends at abstractions and the top tier is simply Ab4 = P1. However, this tier is necessary to further understand the skills involved in expert teaching. An expert teacher must understand both her own learning and her learner's learning at the level of abstractions. In order to link these two, she must be able to form a system of principles.

Audrey, a middle school humanities teacher with a classroom of seventh graders, has a teaching capacity that is within the abstractions tier and extends to include the tier of principles. It turns out that Audrey gets in trouble quite a bit for designing her own curriculum. Though she is in a small progressive public school, her principal doesn't support changing the humanities curriculum from years past (AoC). "You should be teaching children to become readers, writers, and historians," is what Audrey would often hear in staff meetings. During one meeting the principal reprimanded Audrey for creating a small business project for students: "Why do thirteen-year-olds need to learn how to design a small business? You're supposed to be teaching the colonies this time of year! You know, there are really smart people out there who design curricula. You should just follow them and use the teacher guides they create." Audrey, who for years had been vocal about her ambition to motivate students and teach them to become lifelong learners, and to teach them about life in general in a way that they cared about and could connect with (AoST—values), lost her temper and replied, "I see now that this has all just been a big misunderstanding. All of this time I thought I *was* one of those really smart people who could design curricula."

She walked back to her room feeling deflated and angry, but she noted her mood quickly changed when a group of three students ran up to her with their new floor plans (AoST—meta-emotion). "This is where we're putting the research room! Can we go over it with you today? We finally got it." They were about a month into their small business project, and it was just about to get really interesting. Up until this point students had thought they were just having fun creating a kid-friendly business. They had often complained that they never had anyplace to go after school, somewhere they could hang out with their friends, do their homework, eat and drink, get extra help, and pursue their interests and hobbies. The only places they had to go, they felt, were parks and fast-food places, and they felt they were being treated like second-class citizens (AoL—theory of emotion).

Audrey had taken this as a great opportunity to design a project to teach them about the complicated interplay of motivations, decisions, and consequences (AoL—theory of cognition)—not unlike what had taken place in the thirteen colonies during the eighteenth century. So she had told the students they should stop complaining and come up with a better option. With Audrey's support, students formed teams and designed innovative businesses for which they developed complex business plans. Using an American Express small business plan as a template, they created mission and vision statements, schedules of output goals, budgets, and boards of directors. The plan had to be presented to a "loan officer" (Audrey), and once they had received "money," the students had to purchase their location and design the store's floor plan. This may seem like an odd project for seventh graders, but Audrey had carefully planned out each step (AoTP—routines, organization, long-term planning, etc.).

The business plan required students to hone their skills in expository writing, while the development of a board of directors and business practices gave them insight into, among other things, how the three branches of government were formed and continue to function today. Defending their plan required them to practice public speaking and persuasive writing. The financial plan targeted their math skills as well, and several students sought help from their math teacher, who had assisted Audrey in designing that component of the project. Math was again required when students had to purchase their store location and design a floor plan; art was of course involved as well. Students were strongly invested in their plans, since the whole impetus was rooted in addressing their real-life needs. Audrey anticipated that the project would help students better understand the motivations of the American colonists because, like them, the students were motivated by the opportunity to seek a better life and equality and were excited about the financial opportunity of the endeavor (AoL—theory of memory).

After receiving their "loan," the students were also now financially obligated to a higher authority, just like the founders of the colonies were to the king of England. Throughout all of the project workshop sessions Audrey continued her reading, writing, and history classes in a business-as-usual way. Alongside their project time they studied the thirteen colonies, while in math and art they were learning content and skills to help them advance their project. Audrey had wanted the students to decide that they could use information from other classes to help them work through and understand what they felt were their own projects, and in fact that is what they did (AoL—theory of cognition, theory of memory).

Though Audrey's confidence had taken a hit after her morning meeting with the principal, she went forward with the next phase of her plan (AoC—her principal's expectations; AoST—meta-emotion). She announced to the students that their budgets were going to be cut in half. The students had to decide how they were going to react, and they had one week to redesign their plan and present it again to the loan officer. The outcome was fascinating: in an unnerving resemblance to slavery, most of the groups decided to cut their workers' pay and instead offer them food, clothes, and lodging for their labor (AoL—theory of cognition). This made clear to the students the seeds of bondage—a topic Audrey had always struggled to teach because students often couldn't imagine how ordinary people could come to justify slavery, and they had a hard time understanding history as a series of choices made by everyday people, some just like them. Taking the learning experience in this direction was indeed a tricky move, but Audrey felt confident in the relationship she and her students had built throughout the first half of the year (AoI), and she knew that her students would not resent her for designing the project to show them how ordinary people like them might have resorted to slavery. In the absence of such an AoI, this choice could have been emotionally damaging to her students.

Audrey had also planned to begin charging increased interest in the last third of the project, hoping that—much like as had occurred with the American Revolution—the students would begin to start murmuring messages about breaking away from the bank.

This was a difficult project, and for each student to meet these expectations Audrey had to plan carefully, scaffold all lessons, meet student's individual needs (AoTP), and reflect and revise her plans, all while dealing with external forces including her skeptical principal and some anxious parents (AoC). This project wasn't the traditional type of humanities assignment, and that left many of the adults uneasy. However, in the end students were able to use their own goals and motivation to understand the traditional humanities content and skills at a much deeper level (AoL). The integrated connections to math and art helped to support them in their endeavors. While the project didn't exactly help Audrey's relationship with her principal, she felt more fulfilled in her own professional goals, and parent support grew as they saw their children's motivation and confidence in their achievements increase (AoST).

This example of Audrey and her students shows the expert teaching brain in action. The small business project involved a mixture of student and teacher context. The project incorporates AoL because it is student generated and relevant to their developmental age. The business model stemmed from students being frustrated about having no place to go after school. They needed to find—or, in this case, create—businesses that understood and catered to their needs, which required ToM. It also delivered on traditional student-centered lesson goals (e.g., expository and persuasive writing, required curriculum content related to the thirteen colonies, and public speaking). Looking at it from the teaching brain framework, this project tapped into the students' theories of mind, cognition, emotion, and memory—providing a critical "hook" or "trigger" to increase their motivation and understanding.[5] It also incorporated the teacher's context (AoST and AoTP) and her preference for engaging students in real-world interests. The project deepened their exposure to varied perspectives, increased cross-disciplinary collaboration, and offered scaffolding for each student during one-on-one project workshop time.

By exploring the processing that took place in an expert's teaching brain, we see that the most effective interactions happen when teachers utilize their awarenesses to design more accurate ToLBs for their learners. We can better understand why Audrey made the teaching choices she did, and we can even recognize where breakdowns might occur. What was particularly notable is that Audrey formed a ToLB for each student in her class. Her awarenesses occurred both on the individual student level and for the class as a whole. By forming a strong ToLB for not one but multiple students, she exceeded the level of abstract systems and began rising through levels within the principles tier.

Expert Classroom Teacher

This figure depicts teaching within the principles tier of Fischer's Skill Complexity Scale (Figure 7). Refer to Appendix C for an overview of the master systems thinking exhibited in this teaching example.

SENSING* | Comprehensive ToLB Sensing

heightened

AoL: THEORY OF EMOTION
My students "feel like second-class citizens with no kid-friendly place to hang out after school."

heightened

AoC: PRINCIPAL'S EXPECTATIONS
My principal said, "There are really smart people out there who design curricula and I should just follow them!"

heightened

AoST: VALUES
"I know how to design curricula I am one of those really smart people!"

PROCESSING* | Teaching Brain Processing

heightened

AoL: THEORY OF MEMORY
The teacher facilitates a connection between the students' memory of the American colonists and their current experience struggling to find a free place to "hang out."

heightened

AoST: META-EMOTION
The teacher's personal anger and resentment deflated when students ran up in excitement with new project plans.

heightened

AoI: CONNECTION
Given the strong connection with students the teacher felt safe cutting their project budget compelling students to resort to slave owner tactics.

RESPONDING* | Comprehensive ToLB Teaching

heightened

AoL: THEORY OF COGNITION
Empowered students to come up with a better option for their after school interests and design a kid-friendly business.

heightened

AoTP: FLEXIBILTY
Individualized student planning and scaffolding coupled with regular teacher reflection and revision of curricula.

heightened

AoST: PROFESSIONAL GOALS
Designed an interdisciplinary project to stimulate student motivation, meet academic standards, parallel historical events, and challenge students cognitive creativity.

* all 5 awareneses utilized throughout

Figure 10

Additionally, Audrey is cognizant of the external influence that her school principal has on her ability to teach comfortably (AoC). This awareness of context is something that Audrey will have to contribute more effort toward understanding if she expects it to help her truly support her students. All of her students are affected by the school's principal and the culture of the school. If Audrey continues to struggle to find a happy medium, her ToLBs will also suffer. *All* of the awarenesses work toward building more accurate ToLBs; as we build one awareness, it helps us in building another. Struggles in building a specific awareness can also decrease or stunt the development of our teaching brains.

In the next chapter, you will have an opportunity to consider and assess these awarenesses in yourself.

8

YOUR TEACHING BRAIN

We've gone through the theoretical frameworks and in-depth case examples. Now it's time to focus on you and how this connects to seeing yourself as a teacher.

If teaching, like learning, is a developmental skill in all humans, then you might be wondering where you are in this process. While there's no test that you can take that will give you a definitive answer, you can come to a better understanding of where you are as a teacher by being reflective.

Are you a parent focused on supporting your child in the best way possible? Maybe you are a classroom teacher who relies on allegedly tried-and-true lesson plans handed down through the years, or those all-too-familiar prepackaged curricula. If you are a master classroom teacher, do you recognize yourself in some of the examples on these pages?

In this chapter, let's put you front and center. It's time to ask yourself if the concept of the teaching brain makes sense to you, and if so, how it applies to your teaching life. You may need to exit your current paradigm of teaching, and even step outside of your comfort zone. The framework for the teaching brain first and foremost asks you to become aware of yourself as a teacher and as a person, to look closely at the elements that make up your identity.

I've designed five sets of questions that will help you become more aware of the different dimensions of your teaching processes. The questions are intended to create an inventory of the personal and professional characteristics that affect your teaching. This won't leave you with some numerical score or define whether you are a good or bad teacher. Those are the wrong questions. We are all teachers, and each of our teaching brains is at a different stage of development. The questions in this chapter will help you to understand the developmental level of your teaching brain in light of its varied dimensions. If teaching is a system of sensing, processing, and responding, what are your processes? Focusing attention on only your teaching responses won't tell you much about your development of this cognitive skill. You may even have some clear ideas about your thinking process that you've developed over the years, but knowing what aspects to think about and reflect on—the ones that specifically bear on your teaching brain—is a different and more challenging matter. Shedding light on who we are as teachers and how we might become more effective in that pursuit requires a heightened awareness of both the factors that affect our teaching responses, and those that influence what we sense in a teaching-learning context. Basic as it may seem, the first step toward improving your development is recognizing where you are right now. As we saw in the last chapter, teaching requires us to be aware of:

- Yourself as a teacher (AoST)

- The learner (AoL)

- The teacher-learner interaction (AoI)

- Your teaching practice (AoTP)

- The contexts that affect your teaching (AoC)

The following questions were developed to help teachers become more aware of the different dimensions of their teaching brains. Some find it helpful to read through the questions once, without attempting to respond. When you're ready to respond to the questions, keep a record of your responses, and then revisit the questions and your responses a few months from now. Keep in mind that teaching, like learning, is dynamic. Your teaching will change over time—as you, indeed, will continue to change as a full person in multiple contexts. As you respond to these situations, keep in mind that how you respond to a student and situation in an interaction is a mix of many things.

This is not a critique of you as a teacher, nor of your ability or potential to improve your teaching. Rather, it is an approach that will help you become more aware of how you teach, why you teach the way you do, and what to consider as you nurture the development of your teaching brain.

TEACHER AWARENESSES

Awareness of Self as a Teacher

The construct of "awareness of self" is defined in three parts: awareness of private self, awareness of public self, and awareness of perceived self. This survey will ask you to offer feedback on each part. When answering these questions, think about your entire process of teaching, including how you plan, reflect, and respond. Perhaps you're walking in the park and spot a leaf. Before you know it you are teaching your toddler about the different shapes of leaves and when their colors change, or teaching your friend about the different species of trees in the area. Your teaching process doesn't have to resemble a lesson plan such as one a classroom teacher might create when designing a curriculum, but the sight of the leaf set in motion your plan to teach! So answer the questions considering everything *that goes into your teaching, not just what happens in the precise moment you are working with the person you identify as the learner.*

Section I: Private Self-Awareness

Please answer the questions considering your awareness of your private self. This is your awareness of how your identity, personal history and development, values and beliefs, and needs influence you as a teacher. Private self refers to who you are internally.

1. How much do you find yourself thinking about being authentic in your teaching?

Not at all	A little bit	Some	Quite a bit	Very much

2. To what extent do you think about your cultural background in your role as a teacher?

No extent	Slight extent	Moderate extent	Great extent	Very great extent

3. How often are you aware of your personality in your teaching?

Never	Not very often	Somewhat often	Very often	Extremely often

4. In your role as a teacher, to what extent do you find yourself thinking about your personal history?

No extent	Slight extent	Moderate extent	Great extent	Very great extent

5. How much do you find yourself thinking about your development as a teacher?

Not at all	A little bit	Some	Quite a bit	Very much

6. In your role as a teacher, how much do you think about your core values?

Not at all	A little bit	Some	Quite a bit	Very much

7. In your role as a teacher, to what extent do you think about your overall health?

No extent	Slight extent	Moderate extent	Great extent	Very great extent

8. To what extent do you think about your personal fulfillment in teaching?

No extent	Slight extent	Moderate extent	Great extent	Very great extent

Section II: Public Self-Awareness

Please answer the questions considering your awareness of your public self. This refers to your recognition of how your personal characteristics impact the teaching you offer to students. Your public self is what you present to others.

9. How often do you think about the influence of your personality on your teaching?

Never	Not very often	Somewhat often	Very often	Extremely often

10. How much do you think about how your core values shape your teaching?

Not at all	A little bit	Some	Quite a bit	Very much

11. To what extent do you think about how your personal fulfillment impacts your teaching?

No extent	Slight extent	Moderate extent	Great extent	Very great extent

12. To what extent do you think about how your overall health impacts your teaching?

No extent	Slight extent	Moderate extent	Great extent	Very great extent

Section III: Perceived Self-Awareness

Please answer the questions considering your awareness of perceived self. This relates to the extent to which you recognize how your students perceive who you are based on your personal characteristics and behaviors.

13. To what extent do you think about your teaching as being on display?

No extent	Slight extent	Moderate extent	Great extent	Very great extent

14. How often do you think about teaching as involving an audience?

Never	Not very often	Somewhat often	Very often	Extremely often

15. How much do you think about how your students perceive you as a person?

Not at all	A little bit	Some	Quite a bit	Very much

16. To what extent do you try to understand how your students view your cultural background?

No extent	Slight extent	Moderate extent	Great extent	Very great extent

17. How often do you think about how your students view your personality?

Never	Not very often	Somewhat often	Very often	Extremely often

18. To what extent do you think about how your students perceive your core values?

No extent	Slight extent	Moderate extent	Great extent	Very great extent

Your responses to these questions are neither right nor wrong, neither good nor bad. The questions are designed to help you recognize particular traits about yourself and, more specifically, what influences your teaching. What is fascinating and beneficial about measuring self-awareness is that the moment you ask people if they are aware, that exchange in itself tends to make them more aware. The questions you just answered may not have felt like your typical questions on teaching. They were all about you and how you think, a shift that is at the core of this new and developing understanding of teaching. Rather than looking solely at a teacher's outcomes or responses, we need to pay more attention to what a teacher is thinking. What processing is occurring in the teacher's brain when she is engaging in teaching? The self matters greatly, and without this awareness any picture of your teaching will remain incomplete.

Awareness of Learner

Another factor that affects your teaching is the awareness you have of your learner. This is a more common area to consider. However, we rarely ask these questions of anyone outside of the classroom despite the fact that many of us who are not professional teachers engage with learners all the time.

Theory of Mind

- When you observe your learner's behavior, do you consider what has made him or her act in that fashion?

- Have you considered your learner's context when deciphering his or her actions?

Theory of Cognition

- How much do you consider the learner's developmental ability when teaching him or her?

- Do you consider how your learner is understanding the information you have shared with him or her?

- Have you considered whether your learner understands and processes information in the same way that you do?

- How does the student's developmental trajectory impact how you instruct him or her?

Theory of Emotion

- Do you know how your learner is feeling when he or she gets frustrated during a teaching interaction?

- Have you considered how your student's friends affect the way that he or she learns?

- How much do you think about the effect that family dynamics have on the progress of your learner?

Theory of Memory

- Have you ever thought about how far you have to push your learner to recall what you know he or she knows?

- Have you considered what your learner can recall when he or she is working independently, versus what he or she can recall when in a group setting?

- Do you think about how working with friends or family affects your learner's memory?

Having an awareness of your learner means that you recognize the learner's current developmental ability as well as his or her trajectory beyond the immediate teaching interaction. This understanding is about considering the learner's present and future needs as well as all of the variables that affect his or her learning. A learner is not just brain capacity. A learner is made up of a combination of multiple lived experiences and perceptions. In order to have an awareness of your learner, you must recognize his or her specific learning needs in the moment as well as his or her holistic needs, which impact the present and future. As teachers, we often are able to do this by taking the perspective of the student.

Awareness of Teaching Practice

When you have an awareness of your teaching practice, you are reflecting about the thoughts you have (cognition) and the actions you engage in (performance) as a teacher. The thoughts, actions, routines, and decisions are all part of the "stuff" that makes up your teaching. This includes the careful planning you might make for learning to occur—perhaps it's a day at the museum, strategies to create routines such as reading bedtime stories, or ways of forming habits of mind that will encourage lifelong learning. A description of your teaching practices might reflect either tangible actions (routines, organization) or intangible processes (culture, systems thinking). When considering your teaching practice, challenge yourself to explain the thinking behind your actions and how those actions fit into your long- and short-term goals for yourself and your learner. The more you do this, the better you can distinguish between your educational values and your actual teaching practice. A teaching practice is about the teaching that you actually enact and why; it's the "how" mechanism of your teaching, in terms of both your thinking and the performance of that thought.

- What type of culture do you try to create when teaching a difficult concept? Are you part of a culture that you feel prevents you from engaging in thoughtful teaching?

- As you are teaching, are you able to recognize the other influences affecting you and the learner? Are you able to intentionally affect these influences to your advantage?

- How often do you think about the type of person you want your learner to become? Do you find yourself teaching based on the vision you have for who he or she can become?

- When you notice that the way you teach isn't working for your learner, what do you do differently? Do you think the learner should change, or is it your job to shift? Do you find yourself saying, "This is the way I learned it," expecting your learner to do the same?

- How often do you find yourself shifting your teaching practice based on external feedback? What feedback do you attend to most? Least?

- Does your organization sometimes get in the way of your teaching?

- How often do you recognize the unintentional feedback that affects your teaching, such as the weather, the learner's hunger level, recent events in the news, or the amount of sleep you've gotten?

- How do you actively build your skills as a teacher? Are you working toward developing your teaching awarenesses?

Awareness of Interaction

As we've reiterated throughout this book, teaching—unlike learning—is always an interaction. We can't teach without a learner! That means that advancing our knowledge of teaching needs to go beyond the individual teacher.

To know and understand more about teaching, you need to be aware of the interaction between teacher and learner and how it affects each participant in that interaction. This awareness of interaction has been studied and accepted quite broadly in psychotherapy when considering the relationship between patient and therapist, but it has not yet been transferred to education.[1] For teaching this involves looking at your teaching response and gauging how it reflects the connection and relationship between you and the learner. Perhaps you've had a discussion about how you have affected your student's learning and in turn how she has affected you. Or consider an instance where you have focused on a student's needs and are able to describe what exactly you do with that information and how the student then responds— a feedback loop. You make changes in your teaching because of the feedback the student is giving you, and that shift in turn changes the learning experience for the student and the teaching experience for you. This dynamic can flow in either direction: that is, you could choose to give feedback on a student's actions or response, knowing that the student will then make some sort of change. This type of interaction shows your awareness that student and teacher have an impact on each other, reflecting the causal nature of the interaction: the loop is closed, and the individuals shift or change as a result of the interaction.

Dialogue is a somewhat different type of interaction. An awareness of the teacher-learner interaction turns into dialogue when the teacher and student are listening to and affecting each other and when, as the teacher, you are able to clearly describe this dual investment. For example, recall Eva and her father. The father hears Eva say, "This is not my coat; I can't close it." Her father responds to this comment, and then Eva responds in turn. In this back-and-forth, both learner and teacher acknowledge the dialogue and how it dictates the actions that help move the learner forward.

- Are you able to recognize the impact of your teaching on your learner?

- Do you often consider the mutual effect that your learner has on how you teach him or her?

- During the teaching interaction, are you expecting to deliver information to your learner or to connect with him or her?

- Do you see teaching as forming a type of relationship where you and your student are learning to communicate needs and intent?

- If you were to map who is speaking during a teacher-learner interaction, how often are you the one to offer the answer to the learner's question? Do all roads lead back to you? Do you begin and end each feedback loop?

- How often are your teaching responses based on learner feedback? Is that direct feedback or unintentional feedback?

- Consider a recent teaching interaction. Can you say that you got as much out of it as the learner did? Did it cause you to see your teaching differently?

Awareness of Context

Awareness of context is quite a tall order. The context of the teaching interaction can be any attribute or phenomenon that exists outside the student-teacher interaction (physical space, relationships outside of the direct interaction, school philosophy, etc.). For example, becoming aware of one's context might include how the physical space affects your teaching process (planning, implementing, and reflecting). The way you teach your child while you're on the train in front of a bunch of strangers may be different from what you do in the privacy of your own home. The curriculum design of a classroom teacher in a democratic progressive school would be quite different from what that same teacher might use in a school that espouses a strict or "no excuses" model, even if both were ninth-grade world history courses.

Becoming aware of your context also involves being aware of attributes such as the difference between your mood (and your learner's) in the midst of a severe snowstorm and your mood (and your learner's) on a bright sunny day. Your awareness of how external factors impact student learning is extremely important to your ability to adapt and to be effective as a teacher. What physical environments are you most comfortable teaching in? Which ones do you find least comfortable or more challenging?

- Do you notice a change of mood in your learners when the weather changes?

- What outside forces do you recognize as having a large impact on your teaching? Think about factors such as time of day or how hungry you are.

- What relationships outside the teacher-learner interaction often play a role in your ability to teach your student? Do you appreciate this input or push against it?

- How does your philosophy on teaching match up with that which you find most prevalent in your school or community? Do you find that this helps or hinders your ability to teach?

REFLECTIONS FROM TEACHERS ON THE JOB

Actual responses from current classroom teachers who have gone through this process of reflecting upon their awarenesses underscore how unique and singular the development of each of our teaching brains is. My hope in sharing this feedback is that it will offer additional insight as you begin to delve into your individual process. Though these reflections are from a study that I conducted with twenty-three expert classroom teachers, they echo a wide array of experiences from teachers outside of the classroom (parents, children, spouses, supervisors). (See Appendix B for additional details and methods of the study.)

It is noteworthy that teachers said the process of deep reflection on their teaching had an enormous impact on them. Peggy, a remarkable veteran pre-K teacher, talked about feeling that there is a "skill [of teaching] that we've developed which goes underground when people retire and then it's gone." With a laugh, she said that sometimes she feels like she's in some "second goat field at the petting zoo that no one ever goes to visit." She also worried about how much knowledge about teaching is lost because we just don't know the right way to explore it. She wondered what would happen "if we're just honest and say, 'We start here [as teachers], and depending on what happens in your trajectory and what environment you pick, you may end up here.'"

At the end of her interview Peggy asked if she could talk a bit longer about teaching in general. "In talking with you," she said, "it helps place things that are happening for me in my own development as a teacher. It helps give me an affinity to what is actually happening to me. . . . We [teachers] know we're all experiencing something the same, but it's amorphous. What do you do with this knowledge? I like how you [helped me to] unpack this amorphous thing; you know, to get me [to be] more specific [about my teaching]."

Peggy's moment of clarity didn't come as a surprise. Often teachers take on so much because there is an expectation that the ultimate goal is only to support the learner. However, in doing so, what easily gets lost is our connection to our own teaching process. When we reflect on this process, not only are we better able to support our learners, but we also push our skill level as teachers.

Caitlin, a high school teacher in an affluent suburban school that is mostly white, discovered this connection in a rather surprising way. For years she has dedicated herself to her students, who come from various ethnic and socioeconomic backgrounds (in contrast to the school's overall demographics). She is extremely thoughtful about her process and works tirelessly to meet her students' needs. Caitlin's school tracks students by test scores. Her class with the lowest scores has a majority of low-income black and Latino boys. Throughout our interview she often spoke about this group of boys. She spent hours reflecting on how she could best support their success and meet their needs. She discussed various strategies she employed and was adamant that they were just as capable as her advanced placement students. She held them to the same high expectations because she knew they could achieve.

As she spoke about her diverse student population she noted that while she is very cognizant of the impact that her students' cultural backgrounds have on their learning, her own ethnicity was something she never thinks about. Like most teachers trained to think of their students and not themselves, she prided herself on being student-centered. I asked her whether she noticed how she paid attention to her students' ethnicity but didn't think at all about her own. As she began to talk about it more she admitted, "I guess that's true." She struggled to put it into words: "I've no idea what it would be like if I were a minority and I were in the front of the classroom. I'm assuming I would be more aware of it, but I guess I am still aware of it nonetheless. . . . I mean, I'm struggling with thinking about how it really influences my teaching, though I'm sure it does. I just don't think that I've ever been asked that or really ever thought about it." Caitlin had struggled for years about how to best meet the needs of all of her learners. Though she often thought of how to specifically support her young men of color and had been commended for her work, she never quite felt like she had solved that puzzle. By going through the process of in-depth self-reflection, she considered for the first time whether acknowledging various aspects of the self—in this case, her identification with an ethnicity different from that of her students—could broaden her perspective and help her reach a clearer understanding and heightened awareness of the mulitfaceted interactions she had with her students.

Many teachers, administrators, politicians, and parents believe that the best way to improve teaching is to practice student-centered teaching, and I'm well aware of the many feathers that will be ruffled by suggesting otherwise. We hunt for teachers who we believe are "good teachers" and attempt to replicate their *practice* rather than their *process*. But that process is something quite different—and undoubtedly more complex—than what we can witness in a brief moment of engagement or glean from numbers spit out by a computer.

This was an experience that Andrea, a pre-K teacher in the study, knew all too well. Her successful teaching had become so well known within the education community that she'd been approached by publishers, superintendents, and even charter networks. Andrea had been asked to write books, contribute to blogs, and participate in a video series of her teaching. However, she'd turned down most of those offers. Though she believed passionately in supporting teachers and helping them to better understand learners, she did not think any of these avenues would offer that support. "What I have found that people want from me is an anecdotal story and then [for me to say], 'Here's how you can do it,'" she continued, "They want me to formulate. I know there are things that I do that other people can do, but it's hard without seeing it. And part of it is because even though there is a pattern, it's different than a formula. I can talk to you about how I might handle a situation or how I might look at it, but they're not interested in that. They're interested [in] the snapshot that [a] teacher is going to walk away with. I'm really uncomfortable with that because that just limits so much of who I am and what the children need to get."

Every teacher-learner interaction has a different pattern, Andrea pointed out, because the people involved are different, and they all come with different contexts. She said, "So maybe you can help me with this—actually thinking about the things that are the patterns, so [I can] think about it through a different lens." Andrea's struggle is all too common—but it's rarely spoken about and even less often addressed. We all have moments when we feel an interaction is going really well, but we are not always so able to express what's going on when that experience is taking place—the cognitive process. Instead we focus primarily if not exclusively on the observable behaviors and perhaps on the measurable outcomes. In reality, though, those mean little if we have no insight into the process.

Even in talking with experienced classroom teachers who

have spent years thinking deeply about their process, I have found that they too have a hard time going into depth about their teacher selves. Throughout the interviews I did my best to uncover each participant's awareness of him- or herself as a teacher. The power of this process became extremely clear with Sophie, who was no stranger to teaching or to the power of reflection. She had taught in inner-city public schools for more than a decade and had shared her experience broadly. She had published a practice book, authored several articles in notable education journals, participated in guest lectures, been a member of several prominent groups on teacher quality, and even had her own blog on teaching. Through all of this she still teaches full-time in a public middle school. At the time I recruited her for the study, I wasn't aware of her well-known success as a teacher, but she had come very highly recommended from parents who gushed about how marvelously she worked with their children. I knew such accolades were not easily earned. And so I wondered whether this teacher had anything to gain from going through my still-evolving process of self-reflection, and whether I'd really be able to help her uncover anything about her teaching brain.

Sophie was just beginning to talk about herself as a teacher when she happened upon the word "intellectual" in the interview tool. Up to that point she had been very clear and succinct about her answers. I suspect I hadn't thrown anything new at her until then. Now, however, she said, "I think that I'm somebody that is very much an intellectual, but also, that's not the only part of my identity. So at the same time that I say it's important, it's not something that I'm thinking about." She began to work through the distinction: "Sometimes I'm not trying to sound intellectual at all, because depending on the content I might be talking to thirteen-year-olds and I want a concept that is academic, has academic value, but I want it to sound like something that might just pertain to regular everyday life."

I thought it was interesting that though Sophie considered herself to be an intellectual, she was talking about it as if "being intellectual" was only about the way she presented herself, rather than her cognitive perception of herself and her teaching process. She continued, "I'm sort of a regular person and also an intellectual." Here was this fabulous teacher who had received abundant recognition in and out of school, yet she still was hesitant to internally adopt the title of "intellectual." She instead spoke about "intellectual acts" or "intellectual practice" for her students: "My curriculum may have a lot of opportunities and a lot of intellectual merit for kids to develop intellectually."

To uncover more of Sophie's cognitive process of teaching, I pushed her a bit more: "So what are some of the things that you do throughout your teaching process that you would not define as intellectual and others that you would define as intellectual?" Her immediate reaction was to laugh. "Actually, that seems like a hard question," she said. "I've never thought about this before, I guess. . . . Well, I mean, some of the fun that we have in class sometimes" is not intellectual. I was shocked to hear this—such a masterly teacher was questioning whether the moments of fun that she had in class with her students could be considered intellectual time. She had never considered that this fun time is what enabled her to have deep, meaningful relationships and therefore successful teaching interactions with her students. She perhaps never considered that her role as a teacher was first and foremost not myopically putting students at the center but building the context that was needed for them to have meaningful interactions with one another.

Sophie said, "[Sometimes] the student sings the agenda . . . It's just silly. It's not intellectual. It's play. [Other times] a complicated academic concept might [be to] just chuck phrases to students when I'm first introducing it or ask them a question that does not sound like an intellectual question but it may be leaning toward something that we would then name, and in a more formal academic way that may really push them, [like] 'What did you notice about this?'"

Clearly Sophie had considered her teaching practice quite deeply. She knew when to ask a light question and when to follow up with what she called an intellectual question. She had built her classroom culture by adding in humor and silliness and allowing students to do things like sing the agenda. The result was a democratic classroom culture that empowered students by giving them a voice. I asked, "Is there an intellectual component to having designed the activity that way?" Sophie seemed stumped: "Yes, yes, I . . . yeah, there is. I mean, um, you know, I'm always thinking about what my rationale is for using time in class. . . . So I guess I'm always working intellectually . . . Yeah, that makes sense right? Something may seem like it's not intellectual for the students at the moment, but it is for me . . . and planning is always intellectual." This was a eureka moment for Sophie, it seemed—and evidence that this process of evaluating and reflecting on your multiple awarenesses can be powerful.

My goal in doing this study was never really to help the teachers I interviewed—it was mostly to learn from them. When I met Sophie, I thought I would be the one to gain mounds of knowledge by peering into her thought process, and I did. But what this interaction showed me is that I had been shortsighted: in much the way that teachers often have as much to gain from teacher-learner interactions as their students do, reflecting on her process and her teaching brain during the course of this study did in fact help Sophie see herself in another light.

The tunnel-vision focus on one student-centered approach or another is not inherently bad, either in its intentions or in the practices it promotes. Rather, its flawed nature rests in its incompleteness and imbalanced character, which leads to an underdeveloped picture of teaching and learning. Such a lack of perspective allows ample room for, and may even engender, the conflation of overgeneralized theories about students with hard facts about who they are as learners (and there are generally very few such facts available). As a result, teachers are often left frustrated and confused about the reasons they or their students are struggling, and they are unaware of what they might do to redress the situation. Moreover, ignoring the self conflicts head-on with the acknowledgment that teaching is an interaction. A comprehensive awareness of your full teaching process is the key to understanding both successful and unsuccessful interactions with your students, whoever those students may be and wherever you may find yourself in the role of teacher. If we don't understand what is working, what isn't, and where the breakdown is occurring, we can do little to remedy the situation.

9

THE TEACHING BRAIN AND NEXT STEPS FOR EDUCATION REFORM

I always struggled with shy students. For the first few years of teaching I tried every best practice in the book, but for some reason I still couldn't connect with shy students; there just wasn't the same bond I had with other students. I noticed that other teachers did not have the same problem; rather, they had their own struggles with students who I thought were a dream and easy to teach. Finally I realized it was me. Not being shy myself, I found it easy to view being shy as indicating a lack of effort, confusion, or disinterest. I saw that my emotions were getting in the way—frequently I felt that if those students couldn't see how hard I was trying and clearly didn't care, then why should I try so hard to work with them? It took a lot of self-reflection and honesty to recognize that it was my lens, my personal context, clouding the theories I had formed of shy learners. Building the multiple awarenesses necessary to form a comprehensive ToLB takes time, effort, and a lot of communication.

My focus on understanding the teaching brain highlights the processing component of teaching. What is going on in your mind and brain when you teach? The processing is much more complex than just sensing or responding. Still, many of the efforts in education research and policy focus only on how teachers respond—only on those actions that can be observed. Worse, these actions are often measured through student output, removing the teacher from this consideration almost entirely.

Teaching is a dynamic system of sensing, processing, and responding. It's a natural human endeavor that is not unique to the classroom setting. Teaching is a human cognitive skill that develops over the course of our lifespan. And at the heart of it all, teaching is neither an independent act nor merely a tool. It is an interaction between a learner (or many learners) and a teacher. But each one of those understandings runs counter to how we currently think about and evaluate teaching. If we are able to redefine teaching in those terms, there is the potential for a paradigm shift in our education system. A focus on understanding and valuing what happens in the mind of the student has prompted incredible strides in recent decades, yet because teaching is an interaction, we must fully comprehend the mind and brain of both the learner and the teacher. Expanding this knowledge is nothing short of necessary if we are to have a better grasp on what makes for successful teaching-learning interactions.

I've characterized this process as dynamic, meaning that it is always changing and adapting. Where are you as a teacher? Perhaps you have a better idea of that now, but wherever you are, it does not mean that's where you will always be. Indeed, one of my hopes for this book is to inspire teachers of all sorts to take a new look at their teaching selves and reshape how they teach.

Teachers I encounter in my research have often experienced revelations about themselves as we work through the study with them. Sidney admitted that reflecting on her teaching had made her realize that she was not in the classroom to somehow "save" her students, as if she knew what was best for them. She found herself needing to admit that she was different from them. This was extremely difficult for her, because she always wanted to see the world as an egalitarian place. But it was causing her to struggle greatly in her attempts to understand her students' context and to connect with them. She was a Spanish teacher in a school almost exclusively made up of poor children of color, many of whom already spoke Spanish fluently at home. Upon gaining a greater awareness of herself as a teacher, Sidney explained, "My cultural norms are those that my school is trying to correct against. . . . My husband and I always joke that I close the achievement gap by day and then widen it by night."

She went so far as to teach this explicitly to students. As she explained to me: "In some ways, we're definitely trying to prepare students for college by giving them the keys to power." Referring to Lisa Delpit, an eminent education scholar who advocates for student access to the culture of power, Sidney noted that she wanted *all* her students to develop a language of power.[1] But, as she said, "at the same time . . . we're trying to meet students where they are and acknowledge their own cultural norms." Delpit's work also focuses on preparing teachers to be open-minded toward linguistic diversity, arguing that the language of power often runs counter to the language spoken by lower-income students of color.[2] In a sense she was calling attention to the elephant in the room so she and her students could move past it and begin to reshape their interactions from a place of awareness. She felt it was best if she was transparent with them, she said, and "I believe that is best done by acknowledging who you are as a teacher and your own privilege and norms." Her heightened awareness of self as a teacher helped her to improve her multiple awarenesses of her learners, her teaching process, the impact of the contexts in play, and of course the interactions themselves.

In another school, an eighth-grade science teacher described herself as holding the bar high for her students. She felt strongly that there were no excuses for poor performance. She treated her students as "mini adults" and offered a lot of opportunities to review material. She used a mixture of humor and sarcasm to keep them motivated and on task. However, her students' success was quite mixed—only 50 percent were getting a B or higher in her class. She thought this was primarily a reflection of her students' inability or difficulty with her curriculum, as she believed that the way she designed her projects was strong. It didn't bother her so much because she put little stock in grades. It was only the process that mattered to her, and she felt strongly that once students understood the process of the projects she had designed, everything would fall into place and their confusion would dissipate. It didn't matter if the students' first few quarter grades were poor so long as they left the class with a strong holistic understanding.

After an introduction to the concept of the teaching brain, she began to think about her own background as a student. She recalled that before college she had always felt like a mediocre student because she struggled with science and history. She loved English class, but her success there never made her feel accomplished because it came so easily to her. No one in her family pushed grades, and in general school wasn't interesting to her. She didn't feel motivated to do better in science, and she felt her teachers didn't care whether she did well or not. When she arrived at college, her inner student awoke. She felt motivated by her science courses because they seemed relevant to the real world. She loved the material and was fortunate to find professors who let her craft her own intellectual path. These experiences heavily informed her decision to teach science. Her love of science was so strong that she thought it was impossible that she could somehow be doing it wrong. Looking back, however, she began to question her assumptions. She was tough on her students, constantly pushing them, not allowing for their questions, not questioning her own curriculum design, and refusing to accept any excuses for low expectations. Was it because she was afraid they would have an experience similar to the one she had as a middle school student? Did she fear they would not be motivated to learn science, so she thought *she* had to be their motivation—a constant force pushing them to succeed?

The truth she uncovered was that this was indeed her baggage, and it seeped into her teaching. Not having ever been urged to examine closely her context as a teacher, she'd been teaching for years unaware. Like a harmfully overprotective parent, she was so terrified that her students would make the same decisions she had made as a student that she decided to not allow them any choices. Though her actions came from a good place, they were not helping all or even most of her students, nor her ability to develop as a teacher. She had stayed securely within her comfort zone, rather than becoming aware of the multiple factors that all play a role in successful, effective teaching.

The more aware you are, the more you can attune to your own skill level as a teacher. This is not intended to be criticism; rather, it's a method of staying connected to your development as a teacher. Remember the skill complexity scale? Keep in mind that this is not an achievement scale and it's not linear. It is a tool that helps us understand how the cognitive skill of teaching develops over time in all humans. Some of our teaching ability comes naturally, but like other cognitive skills, it takes a lot of time and specific honing to reach peak levels of performance. In this framework, the peak is the maximum capacity of the human brain to teach—it is *not* how we would all define the "best" teaching. Because teaching is an interaction, "best" depends entirely on the context and the expectations of those involved in that particular circumstance. You want to consider where you presently fall along the complexity scale and where you would like to be. It's when those two match each other that you will more likely achieve teaching success. Like any other relationship, it's about the best fit.

A NEW DEFINITION OF SUCCESSFUL TEACHING IN OUR SCHOOLS

The concept of the teaching brain suggests that we can support teachers by valuing this "better fit" approach to teaching and education in general. Quite intentionally I am not outlining a specific or rigid set of "best practices" that define an expert teacher; nor offering a checklist that defines "good teaching." Rather, the concept of the teaching brain asks that each individual teacher assume responsibility for his or her teaching. This goes back to the essential quality of any dynamic system: variability, especially variability in context. The teaching brain is an empowering, flexible framework that can enable you to evaluate where you are now and help you lay out a path for where you want to go. That is success: finding a match between where you are and where you want to be.

If you're a classroom teacher, the concept of the teaching brain also helps you to determine where your school expects you to be and how you are meeting those expectations and the school's needs. A school dedicated to experiential learning where students are taught in outdoor classrooms may believe that leadership means achieving a sense of self-confidence. In that environment, teaching success may be linked to supporting each individual student in learning outdoor skills to overcome their fears.

By contrast, schools that support a "no excuses" policy seem to define success solely on the basis of student outcomes as measured by standardized test scores. In these cases, successful teachers look at students not as individuals but as groups compared to a mythical average.[3] As we've seen, students are not only individuals but individuals who vary within themselves and in different contexts. Their skills, performance, and learning continually shift depending on the context, the presence or absence of scaffolding, and the dynamics of their interactions with their teachers.

Rather than debating which type of school is best and forcing adoption of those practices, we should be evaluating the fit between teachers and schools. Success is relative, and it should be seen no differently when it comes to our students, our teachers, and our schools. It is based on the values and expectations of the school and the individuals within the teaching-learning interaction. If your school's values directly conflict with your own, as a person or more specifically as a teacher, it is unlikely that you will achieve success with your students in that environment.

Ask yourself, "Where does my school expect me to be? Do these goals match up with where I want to be?" If so, you can carve a path to be successful. If not, then you might ask yourself whether you are in the right place. Are you at a developmental level below where you are expected to be? If so, is there a system in place that gives you guidance for how to set and achieve your teaching goals? If you're at a developmental level above the school's expectations, you might ask yourself whether you agree with the values and vision of the school.

You might be wondering what the teaching brain framework can do for you if you are not a classroom teacher. Recognizing your teaching brain is about helping you to understand how you can form better interactions with the learners in your life. Our children are our most obvious learners, and we struggle to understand those interactions every day. But we are also teachers as we relate to coworkers, siblings, spouses, and even parents. We all take on the role of teacher at various moments, either during a single conversation or over an extended period of time. Do you ever find yourself in the same cycle with your spouse over and over again, trying to teach him the same thing? Have you found yourself saying, "I've told you this a hundred times"? Well, if you don't understand the concept of the teaching brain, you'll likely go through the cycle a thousand more times and still wind up with the same results, getting nowhere. Understanding your teaching brain will allow you to see that the theory you have formed about your spouse (the learner) may be incorrect. Perhaps becoming more aware of yourself as a teacher, and/or of the outside contexts that are exerting pressure on the interaction, will lead to a more accurate perception of the challenges before you.

Whatever the case, heightening your awareness in the various areas described in this book (AoL, AoST, AoI, AoTP, AoC) will help you engage in the brain processing that can improve your ability to sense from your learner, process that information, and develop a more effective response.

WHAT THE SCIENCE TELLS US AND WHAT IT DOESN'T

The narrow conceptions of teaching we commonly employ focus primarily on one of the products of teaching: learning. This model of teaching suggests that the solution to unsuccessful knowledge transmission is to modify the input (the teacher) in order to improve the desired output (student learning). These ideas are borne out in many of the education reforms we see today, including the use of short-term teachers in charter schools, teacher training scorecards, "flipped" classrooms (where instruction—usually lectures—happens online and homework happens in the classroom), and school improvement plans calling for "talented" teachers. But studies of development show us that human teaching is much more complicated than these input-output models, which originated in early research on animal behaviorism.

Such so-called solutions fail to consider teaching as an interactive system in which at least two individuals with complex, dynamic brains are engaged. Without this perspective, it is no wonder that current research and policy focus almost exclusively on learning processes. While research has increased our knowledge of teacher cognition, our recognition and understanding of the processes of the teaching brain and its complicated interactions with the learning brain are still in their infancy.

Unfortunately teaching is still most commonly defined as a tool for learning, and the methods used to conduct research on the effectiveness of a tool are quite different from those that attempt to understand a human interaction or a developmental skill.

The lack of information on the teaching-learning interaction is in part a result of technological constraints. Traditionally, research in cognitive neuroscience works with the brain of only one participant at a time, rendering it impossible to collect the same level of data from multiple people in an interaction. However, exciting new neuroimaging methods are being employed to observe the neural activity of interacting brains in the emerging field of two-person neuroscience (2PN).

In an early 2PN study, Yun and colleagues used simultaneous electroencephalography to measure synchrony in the brain activity of two people in a face-to-face interaction. They observed that across pairs of individuals, interpersonal interactions enhanced activity in the precuneus and right posterior middle temporal area—two regions considered to be a part of ToM networks (implicated in social functions such as empathy). The researchers also found that when participants synchronized their bodily movements, they showed corresponding changes in ventromedial prefrontal cortex activity (a brain region involved in forming representations of the self and social/emotional attitudes).[4] The implications of these findings for understanding teaching are profound. Could a more successful interaction be reflected in neural synchrony?

In a recent study that begins to address these questions, Holper and colleagues conducted the first simultaneous neuroimaging study of teacher-student interactions.[5] This 2PN study utilized wireless functional near-infrared spectroscopy (fNIRS) to record prefrontal brain activity while teachers used the Socratic method to help students solve a math problem. Researchers found that during sessions where students arrived at the correct answer, teachers' and students' brain activity "danced at the same pace." Could neural synchrony be a precursor to successful learning?

These methods are finally allowing us to turn away from

simplistic studies of concrete responses and the products of teaching and instead to focus on the complex processes that occur during the interactions of teaching and learning brains. We need to not only increase our understanding of the teacher-learner interaction but also acquire more information on the teaching brain, equivalent to our knowledge of the learning brain.

RESTORING FAITH IN OUR POWER TO FIND SOLUTIONS

Our understanding of how humans learn has evolved from simple concepts of an input-output model to cognitive ladders of learning development and finally complex context-dependent systems. The holistic "learning brain" approach to developmental research is modeled on the principles of dynamic systems and has been studied using a variety of modalities, from conceptual model building to brain scans. A similar trajectory exists in our understanding of teaching, moving from early behaviorist perspectives to cognitive approaches that focus on teacher thinking as well as teachers' pedagogical and content knowledge. The behaviorists focused on observing teaching behaviors, using objective measures to determine the feedback and response between teacher and learner. The cognitivists created models of the teacher's mind, developing constructs such as ToM and PCK. Now teaching research is just beginning to embark on its next phase of exploration. This work borrows from the dynamic understanding of the learning brain to investigate teaching in a holistic fashion that simultaneously examines the what, how, and why of teaching. The stage is set for further research into the teaching brain using the full spectrum of educational and cognitive psychology and brain imaging research tools that will help us map the next chapter in our understanding of human teaching.

Without a full understanding of the teacher, we are ignoring the fact that teaching is an interaction. Previous research on teacher cognition has given us insight into the mind of a classroom teacher. This book is an attempt to lay out a much-needed new framework for how we think about teaching. Only if we support and encourage teachers to recognize themselves as an independent part within the larger teacher-learner interaction will they understand how their lens and intentions affect that system.[6] Rather than blaming a student's lack of success exclusively on the learner or exclusively on the teacher, everyone affected can become aware that a teaching response was based on an incorrect theory of the learner's brain, a lack of awareness of the context, or perhaps problems with the actual teaching practice. And *then* a clearer picture of a better way can surface. Like most anything that truly matters, it's more complicated than a simple black-or-white solution.

The teaching brain framework allows us to explore teaching along the life span, from birth through adulthood, and enables us to embrace a holistic view of teaching. Once we find out where we fall on the skill complexity scale for teaching, we can have a better sense of how we fit into our own teaching contexts. The best fit can be demonstrated by the synchrony and flow that occurs between all the players. Though on the surface this seems contrary to a student-centered approach, in fact it offers a more comprehensive approach to student success—an approach that allows for everyone to win. Instead of a one-size-fits-all teaching approach, which assumes that all learners are the same, the teaching brain model acknowledges the need to understand both the teacher and the student as well as their context and interaction. Student-centered approaches that focus only on student outcomes as a measure of teacher success can easily hide the variability across a teacher's skill development and the enormous difference that can be found from one teacher-learner interaction to the next. This can leave schools, teachers, learners, and policy makers at a loss for how to improve teaching. The teaching brain model, I suggest, can serve as a source for solutions.

Rather than continuing to drown in the failure we feel exists in our school systems, we should gain solace and renewed faith by remembering that at the heart of education, all humans are teachers. We have it within our power to find the solution to our educational woes if we finally ask the right question: What is teaching? We can begin to answer this question by understanding how this cognitive and emotional skill develops over time, and specifically how it develops in expert teachers. Science can play an important role here, helping to uncover the physiological and brain processes teachers engage in during a teaching-learning interaction. By asking this precise question, the worlds of science and education can join together to finally demystify what makes a natural teacher—someone who can say, "You know when it's working; you can just feel the synergy."

APPENDIX A

ABBREVIATIONS

AoC	awareness of context
AoI	awareness of interaction
AoL	awareness of learner
AoST	awareness of self as a teacher
AoTP	awareness of teaching practice
CLASS	Classroom Assessment Scoring System
ELA	English language arts
NCLB	No Child Left Behind
PCK	pedagogical content knowledge
TFA	Teach for America
ToLB	theory of the learner's brain
ToM	theory of mind
UDL	Universal Design for Learning
UTC	unified theory of cognition
ZPD	zone of proximal development

METHODS USED TO INVESTIGATE THE COGNITIVE PROCESSES OF THE TEACHING BRAIN

The design of the interviews referenced throughout this book was premised on supporting teachers to describe both the cognitive processes they use when teaching and the entirety of their teaching process—from instructional planning through implementation and reflection. Participants were asked to respond to an interview question both in free-response and scaffolded interview formats.

Since I was looking for in-depth analyses of participant response and behaviors within a specified time period, this design choice draws upon microgenetic interview methods, as well as an understanding of the developmental processes as they occur.[1] For this reason, the interview protocol also utilized and built upon the Self in Relationships tool (SiR), which enables individuals to see themselves in context with their relationships.

Since self-descriptions are often difficult to measure and assess reliably—one's understanding of self is embedded within personal context and relationships—we adapted this microgenetic interview method (SiR) to include a question related to the individual's cognitive processes of teaching, in both a low- and high-support environment.[2] Typically in a microgenetic study participants repeat the same problem-solving task within one session in order to assess intra-individual variability.[3] The variability is derived from how the participant modifies her answers as she moves forward with completing the task. However, it has also been shown that a modification in the interview protocol itself can elicit significantly more complex responses from participants.[4] Considering that this study aimed to discover teachers' cognitive processes throughout their teaching, I adapted the interview context to encourage a shift in how the teachers performed the task.

The task remained constant, but participants were given a high-support tool to utilize in crafting their responses. The intent with this study design was to encourage developmental change in participants. We predicted that in a high-support condition, the teachers would be able to answer the interview question with a significantly more complex response.[5] This would then enable us to observe peak performance of their cognitive skill of teaching.

SELF-IN-RELATION-TO-TEACHING (SIR2T)

The SiR method was originally used to explore adolescents' perspectives of themselves in realtion to others.[6] The approach is a complex, microgentic interview and assessment that involves multiple steps. Research findings on teaching have not produced a developmental skill theory of the teaching brain. By organizing the cognitive processes of the teacher's brain (as shared by the teacher) during his or her process of teaching, we intended to design a new model to explore the teaching brain in future studies.

While there is no skill theory specific to teaching, we hypothesized that the cognitive processses involved in teaching could be comparable to learning in that they are both dynamic and context dependent.[7] Teachers' self-descriptions are often confounded with student performance and therefore difficult to tease out.[8] Therefore, both low- and high-support situations were included in the modifed SiR interview protocol (termed SiR2T, or

Self-in-Relation-to-Teaching) to uncover teacher thinking more effectively. Additionally, with both low- and high-support settings teachers are more likely to come to understand the abstract systems involved in their teaching process and then be able to describe them.

In the low-support situation, teachers participated in an open-ended writing exercise in response to the question "What are you focusing your mind on throughout the process of teaching?" The question focused on the teacher's process specifically so that teachers would not merely discuss when they were physically in front of the classroom or with students while teaching. Significant research on teacher thinking has demonstrated that planning and reflection are important aspects of teacher cognition.[9] During the low-support condition, teachers simply responded to the question in free form via e-mail. The intent of this portion of the interview was to record a teacher's first reactions to the concept of his or her cognitive processes without directing a teacher toward addressing specific topics. By design, this low-support situation would yield functional performance and serve as a priming exercise for participants as they considered the full spectrum of their teaching process (from planning through reflection). By priming them to consider this full spectrum, we postulate that teachers would be more situated within their context and, therefore, more likely to provide robust descriptions of their self-understanding.[10]

The high-support portion of the interview was comprised of the SiR2T self-diagram task and clarifying questions, which consisted of an initial speculative set of teaching brain characteristics drawn from literature on cognitive science,[11] developmental and cognitive psychology,[12] biology,[13] and teacher education.[14] The hypothetical teaching brain characteristics were broken down into categories of self, personal context, skills, and external influences.

APPENDIX C

TYPES OF TEACHING ABILITY DEVELOPED ACROSS THE LIFE SPAN

This table[1] shows the shift in teaching emphasis at different ages. The age gaps indicate neither an absence nor a sudden emergence in teaching development at any specific age. Instead the stages shown are intended to highlight sample teaching practices across the lifespan, culminating with an additional phase of expert teaching to reflect the advanced development in adults who actively cultivate their teaching brains.

Age/ Phase of Teacher	Type of Teaching	Description of Teaching Strategy
1 year old	**Proto-teaching:** Recognizing a knowledge gap, the teacher corrects the learner to close it.	In a study where the experimenter tries to put a triangular-shaped object in a round hole, the teacher points to the correct hole.[2]
3 years old	**Demonstration:** A physical demonstration without the learner's involvement.	When asked to teach another child to play a game, the child is able to do so through demonstration but without much explanation. The teacher also corrects the learner's errors during the game, but the correction is not followed by any explanation.[3]

5 years old	**Explanation:** Information is explained and often accompanied by a physical demonstration.	Here the *dominant* teaching strategy is explanation. If the learner makes a mistake, the teacher will explain the rule again in a shortened version or through demonstration. The teacher can monitor the learning process and adapt his or her explanation and demonstration based on the learner's progress.[4]
7 years old	**Contingent Teaching:** This is scaffolded teaching, which is contingent on the learner's progress.	The teacher's attention is focused on whether the learner becomes competent in the task or frustrated. As a rule for contingent teaching: if the learner succeeds, offer less help in your next intervention. If the learner fails, take over more control when intervening.[5]
9–11 years old	**Strategic Advice & Choice:** The teacher helps the learner think through the pros and cons of his or her choices.	While quite similar to teachers at age 7, at this age a teacher might also offer strategic advice regarding alternative choices available to the learner.[6]
Adult	**Mental Model:** The teacher forms assumptions and theories about the mind of the learner and how learning is happening; the teacher acknowledges that those assumptions inform his or her teaching.	There are two types of mental models: 1. Espoused models are culled from the ways people speak about their teaching. 2. In-action models are inferred from actual teaching.[7]

Strauss's table ends there. Based on my analyses of the teaching brain, I would add the following stage:

Expert Teacher	**Master Systems Thinker:** The teacher has mastered systems thinking and responds based on a comprehensive understanding of all five awarenesses.	Expert teachers are able to recognize and manage the mutuality of the teacher, the student, and the larger interactive system. This systems thinking supports expert teachers to adapt constantly to changing environments.

ACKNOWLEDGMENTS

My inspiration is, and has always been, my students. My vision and drive to explore the process of teaching grew out of the joy I derived from being with them. Although I cannot thank each of you personally, please know that you have meant everything to me.

I am forever grateful to Michelle (Billie) Fitzpatrick for agreeing to be my partner in this endeavor. From the moment we first met in a "Mind, Brain, and Education" class at Harvard, her never-ending support for the teaching brain has been invaluable. Billie's expertise in writing and her calming nature talked me off many a ledge when I thought I couldn't possibly write a single word more. She has not only been a writing companion and thought partner, she also has become a very dear friend.

I owe a very special thanks to The New Press. Years ago I happened upon its booth at the annual meeting of the American Educational Research Association. Until that moment I hadn't realized that its list was home to many of my staple teaching references. It was at that same conference that I met Ellen Reeves, then The New Press's education editor. She invited me to join the education committee, a group charged with advising the press on the substance and character of its education publishing program.

I was moved by the house's commitment to valuing the perspectives of actual classroom teachers. When I began the process of finding the perfect publisher for this book, I contacted Diane Wachtell, the executive director, who introduced me to my editor, Tara Grove. Tara spent countless hours shaping and sharpening this book. I suspect it is rare to find an editor willing to give so much to help an author express her message clearly. Never was I asked to simplify the complexity of the ideas on these pages (even when the concepts risked demanding some heavy lifting by readers); rather, they insisted I stay true to my vision. The New Press has truly been a gift.

This book, like most, is the result of many relationships and many years of teaching, thinking, and writing. Early in my career I received an Action Research Fellowship from the Teachers Network Policy Institute. Christopher Clark helped open my eyes to the world of teacher research. Chris has continued to be an amazing mentor through all of the twists and turns of my career over the years. An additional thanks goes to Klaus Bornemann, whose unfailing confidence in me was a miracle in dark times.

About halfway through my teaching career, I met Robert Cohen, former chair of teaching and learning at New York University's Steinhardt School of Education, my alma mater. When he offered a helping hand, I'm sure he never could have predicted that he was getting himself into years' worth of planning together, designing projects, and co-teaching. He was critical to the process of publishing my first article and played a key role in my effort to represent teachers' voices in academia. Robby's support was unwavering and his respect for classroom teachers is uplifting.

NYU is also where I met my dear friend Deborah Meier, who has always challenged me to think through my ideas and consider the alternatives. Deb never lets my thinking get lazy yet is always happy to let me cozy up in her home to recover from life's challenges. It was she who introduced me to Linda Nathan, an extraordinary educator who graciously put me up when I first arrived in Cambridge. Linda has offered her support, in work and in life, ever since.

For many reasons, I never thought I'd be in a doctoral program at Harvard. Meira Levinson at the Harvard Graduate School of Education (HGSE) convinced me that I was meant to be at HGSE. She said, "If your work existed when I was a classroom teacher, I never would have left. Come here and let us support you in doing what I know you can." It was the first time someone truly understood my vision. Never before had I been given such support and encouragement to do what my mind was capable of. I will be forever indebted. Meira connected me to the extraordinary librarian Carla Lillvik, who answers each and every one of my questions immediately and with thoughtful grace.

HGSE has provided a playland for my brain. It is here that I met my adviser and friend Kurt Fischer. Though I was new to the mind, brain, and education discipline, he was certain it was the right home for me and he couldn't have been more right. Kurt tirelessly supports my ideas; gives me room to explore, fail, and revise; and never hinders me with arbitrary rules. My co-adviser and friend Tina Grotzer has been a consistent advocate for my work. Tina performs magic—asking the right questions and making the right edits until she finds the diamond in the rough. Through the tears and the laughter, Kurt and Tina have seen my greatest fears and joys—and they keep coming back for more!

I'd also like to thank some of the original HGSE administrators and the Harvard Initiative for Learning and Teaching, who were a great support to me in my early years as a doctoral candidate. Shu-Ling Chen, assistant dean of the EdD program, fully supported my efforts to apply for the esteemed Harvard Initiative for Learning and Teaching (HILT) grant. Associate dean Matthew Miller, along with academic dean Hirohito Yoshikawa, worked tirelessly to help me revise many, many drafts of my grant proposal. Many thanks to HILT for rewarding these efforts and for supporting my first studies on the teaching brain. Together with the support of then-dean Kathleen McCartney and the HILT team, I was on my way to mapping the teaching brain.

This book and the studies that have been part of it would have been a lot more painful without the help of my fellow MBE doctoral friends and colleagues Bryan Mascio, S. Lynneth Solis, Courtney Pollack, and Laura Edwards.

I'd like also to thank Antonio Battro and Sidney Strauss, both exceptional experts on teaching, cognition, and the brain. They first welcomed me to the 7th Course of the International School on Mind, Brain, and Education (ISMBE), hosted at the breathtaking Ettore Majorana Centre for Scientific Culture in Erice, Sicily, to share my work on the teaching brain. We have become great friends since then and they continue to mentor and drive my work.

Conventions of anonymity prevent me from naming the many teachers and school administrators whose assistance have been critical to my work. I am deeply grateful for their participation and continued support. I only hope that I have done justice to all that they have contributed.

Of course none of this would be possible without my parents and sisters, who have made me who I am. Loving and always critical, they have taught me to seek happiness, search for answers and never settle for just enough. They believe we always find a way to make it work. To my exceptional nephews, I thank you for the countless teaching and learning moments, the joy and laughter, and for filling the void when I no longer had a classroom. You are how I know I still have the teaching disease.

Last but undoubtedly not least, I would like to thank my very best friend, Devin. Few have ever understood how after sixteen years we are still just as attached at the hip as when we first met. From my very first teaching position to my most recent job, he has helped me brainstorm projects, debate over student work,

and plan career moves. He read every chapter of this book, gave me multiple pep talks, and wiped many tears, and through it all he cooked, cleaned, and brought me to bed. All of what you read, and all of me, could never exist without him.

NOTES

Introduction

1. Liz Riggs, "Why Do Teachers Quit?," *The Atlantic*, October 18, 2013, www.theatlantic.com/education/archive/2013/10/why-do-teachers -quit/280699/.

2. Tovah P. Klein, *How Toddlers Thrive: What Parents Can Do Today for Children Ages 2–5 to Plant the Seeds of Lifelong Success* (New York: Touchstone, 2014); Kim John Payne with Lisa Ross, *Simplicity Parenting* (New York, Ballantine, 2010); Harley Rotbart, *No Regrets Parenting* (Kansas City, MO: Andrews McMeel, 2012).

3. Elizabeth Green, "Building a Better Teacher," *New York Times Magazine*, March 2, 2010.

1. The Theories That Have Led Us Astray

1. Tim M. Caro and Marc D. Hauser, "Is There Teaching in Nonhuman Animals?," *Quarterly Review of Biology* 67, no. 2 (1992): 151–74.

2. Nichola J. Raihani and Amanda R. Ridley, "Experimental Evidence for Teaching in Wild Pied Babblers," *Animal Behaviour* 75, no. 1 (2008): 3–11, doi:10.1016/j.anbehav.2007.07.024.

3. S.A. West, A.S. Griffin, and A. Gardner, "Social Semantics: Altruism, Cooperation, Mutualism, Strong Reciprocity and Group Selection," *Journal of Evolutionary Biology* 20, no. 2 (2007): 415–32, doi:10.1111/j.1420-9101.2006.01258.x.

4. Nigel R. Franks and Tom Richardson, "Teaching in Tandem-Running Ants," *Nature* 439, no. 7073 (2006): 153, doi:10.1038/439153a; Raihani and Ridley, "Experimental Evidence"; Alex Thornton and Katherine McAuliffe, "Teaching in Wild Meerkats," *Science* 313, no. 5784 (2006): 227–29, doi:10.1126/science.1128727; Dario Maestripieri, "Maternal Encouragement in Nonhuman Primates and the Question of Animal Teaching," *Human Nature* 6, no. 4 (1995): 361–78.

5. B.F. Skinner, "Why We Need Teaching Machines," *Harvard Educational Review* 31, no. 4 (1961): 377–98.

6. A.A. Lumsdaine, "Teaching Machines and Self-Instructional Materials," *Audio-Visual Communication Review* 7, no. 3 (1959): 163–81.

7. J.H. Ward Jr. and K.J. Davis, "Artificial Intelligence and Self-Organizing Systems: Teaching a Digital Computer to Assist in Making Decisions," *Proceedings of the 1962 ACM National Conference on Digest of Technical Papers* (1962): 11, doi:10.1145/800198.806077.

8. Claire M. Fletcher-Flinn and Breon Gravatt, "The Efficacy of Computer Assisted Instruction (CAI): A Meta-Analysis," *Journal of Educational Computing Research*, 12, no. 3 (1995): 219–41.

9. Andrew R. Molnar, "Computers in Education: A Historical Perspective of the Unfinished Task," *Technological Horizons in Education* 18, no. 4 (1990): 80–83; Richard P. Niemiec and Herbert J. Walberg, "From Teaching Machines to Microcomputers: Some Milestones in the History of Computer-Based Instruction," *Journal of Research on Computing in Education*, 21, no. 3 (1989): 263–76, doi:10.1080/0888650 4.1989.10781877.

10. Eric Jensen and Gary Johnson, *The Learning Brain* (San Diego: Brain Store, 1994).

11. Ann Gordon and Kathryn Browne, "Developmental and Learning Theories," in *Beginnings and Beyond: Foundations in Early Childhood Education*, 9th ed. (Belmont, CA: Cengage Advantage Books, 2013), 99.

12. Stacey T. Lutz and William G. Huitt, "Connecting Cognitive Development and Constructivism: Implications from Theory for Instruction and Assessment," *Constructivism in the Human Sciences* 9, no. 1 (2004): 67–90.

13. John Dewey, *Experience and Education: The 60th Anniversary Edition* (New York: Kappa Delta Pi, 1998); Maria Montessori, *The*

Child, Society and the World: Unpublished Speeches and Writings, ed. Günter Schulz-Benesch and trans. Caroline Juler and Heather Yesson (Oxford, UK: Clio Press, 1946); David A. Kolb and Robert Fry, "Toward an Applied Theory of Experiential Learning," in *Theories of Group Process*, ed. Cary L. Cooper (London: John Wiley, 1975); Kurt W. Fischer and Arlyne Lazerson, "Research: Brain Spurts and Piagetian Periods," *Educational Leadership* 41, no. 5 (1984): 70.

14. David A. Sousa, *Mind, Brain, and Education: Neuroscience Implications for the Classroom* (Bloomington, IN: Solution Tree Press, 2010).

15. Caro and Hauser, "Is There Teaching in Nonhuman Animals?"

16. David Premack, "The Aesthetic Basis of Pedagogy," in *Cognition and the Symbolic Processes: Applied and Ecological Perspectives*, ed. Robert R. Hoffman and David Stuart Palermo (Hillsdale, NJ: Lawrence Erlbaum Associates, 1991), 545.

17. Sidney Strauss, Margalit Ziv, and Adi Stein, "Teaching as a Natural Cognition and Its Relations to Preschoolers' Developing Theory of Mind," *Cognitive Development* 17, no. 3–4 (2002): 1473–87, doi:10.1016/S0885-2014(02)00128-4.

18. Sidney Strauss, "Teaching as a Natural Cognitive Ability: Implications for Classroom Practice and Teacher Education," in *Developmental Psychology and Social Change: Research, History, and Policy*, ed. David B. Pillemer and Sheldon H. White (New York: Cambridge University Press, 2005), 368–89.

19. Antonio M. Battro, "The Teaching Brain," *Mind, Brain, and Education* 4, no. 1 (2010): 28–33, doi:10.1111/j.1751-228X.2009.01080.x; C.M. Heyes, "Theory of Mind in Nonhuman Primates," *Behavioral and Brain Sciences* 21, no. 1 (1998): 101–14; Derek C. Penn and Daniel J. Povinelli, "On the Lack of Evidence That Non-human Animals Possess Anything Remotely Resembling a 'Theory of Mind,'" *Philosophical Transactions of the Royal Society of London, Series B* 362, no. 1480 (2007): 731–44, doi:10.1098/rstb.2006.2023; David Premack and Ann James Premack, "Why Animals Lack Pedagogy and Some Cultures Have More of It Than Others," in *The Handbook of Education and Human Development: New Models of Learning, Teaching and Schooling*, ed. David R. Olson and Nancy Torrance (Cambridge, MA: Blackwell, 1996), 302–44; Strauss et al., "Teaching as a Natural Cognition"; Alex

Thornton and Nichola J. Raihani, "The Evolution of Teaching," *Animal Behaviour* 75, no. 6 (2008): 1823–36, doi:10.1016/j.anbehav.2007.12.014.

20. Sidney Strauss and Margalit Ziv, "Teaching Is a Natural Cognitive Ability for Humans," *Mind, Brain, and Education* 6, no. 4 (2012): 186–96, doi:10.1111/j.1751-228X.2012.01156.x; Thornton and McAuliffe, "Teaching in Wild Meerkats."

21. Allen Pearson, *The Teacher: Theory and Practice in Teacher Education* (New York: Routledge, 1989).

22. Doug Frye and Margalit Ziv, "Teaching and Learning as Intentional Activities," in *The Development of Social Cognition and Communication*, ed. Bruce D. Homer and Catherine S. Tamis-LeMonda (Mahwah, NJ: Lawrence Erlbaum Associates, 2005), 231–58.

23. Ulf Liszkowski, Malinda Carpenter, Tricia Striano, and Michael Tomasello, "12- and 18-Month-Olds Point to Provide Information for Others," *Journal of Cognition and Development* 7, no. 2 (2006): 173–87, doi:10.1207/s15327647jcd0702_2.

24. Strauss and Ziv, "Teaching Is a Natural Cognitive Ability."

25. David Wood, Heather Wood, Shaaron Ainsworth, and Claire O'Malley, "On Becoming a Tutor: Toward an Ontogenetic Model," *Cognition and Instruction* 13, no. 4 (1995): 565–81, doi:10.1207/s1532690xci1304_7.

26. James Garbarino, "The Impact of Anticipated Reward Upon Cross-Age Tutoring," *Journal of Personality and Social Psychology* 32, no. 3 (1975): 421–28; Russell J. Ludeke and Willard W. Hartup, "Teaching Behaviors of 9- and 11-Year-Old Girls in Mixed-Age and Same-Age Dyads," *Journal of Educational Psychology* 75, no. 6 (1983): 908–14, doi:10.1037/0022-0663.75.6.908.

27. Caro and Hauser, "Is There Teaching in Nonhuman Animals?"

28. Robert J. Havighurst, *Human Development and Education*, 1st ed. (New York: Longmans, Green, 1953).

29. Christopher Clark and Magdalene Lampert, "The Study of Teacher Thinking: Implications for Teacher Education," *Journal of Teacher Education* 37, no. 5 (1986): 27–31, doi:10.1177/002248718603700506; Strauss, "Teaching as a Natural Cognitive Ability."

30. Ryuta Aoki, Tsukasa Funane, and Hideaki Koizumi, "Brain Science of Ethics: Present Status and the Future," *Mind, Brain, and Education* 4, no. 4 (2010): 188–95, doi:10.1111/j.1751-228X.2010.01098.x.

2. Cookie-Cutter Solutions and Other Missteps of Education Reform

1. James W. Fraser, *The School in the United States: A Documentary History* (New York: Routledge, 2009).

2. Motoko Rich, "At Charter Schools, Short Careers by Choice," *New York Times*, August 26, 2013.

3. David Wood, Heather Wood, and David Middleton, "An Experimental Evaluation of Four Face-to-Face Teaching Strategies," *International Journal of Behavioral Development* 1, no. 2 (1978): 131–47, doi:10.1177/016502547800100203.

4. Sidney Strauss and Margalit Ziv, "Teaching Is a Natural Cognitive Ability for Humans," *Mind, Brain, and Education* 6, no. 4 (2012): 190, doi:10.1111/j.1751-228X.2012.01156.x.

5. Orly Haim, Sidney Strauss, and Dorit Ravid, "Relations Between EFL Teachers' Formal Knowledge of Grammar and Their In-Action Mental Models of Children's Minds and Learning," *Teaching and Teacher Education* 20, no. 8 (2004): 861–80; Miriam Mevorach and Sidney Strauss, "Teacher Educators' In-Action Mental Models in Different Teaching Situations," *Teachers and Teaching: Theory and Practice*, 18, no. 1 (2012): 25–41, doi:10.1080/13540602.2011.622551; Sidney Strauss, Dorit Ravid, and Nicole Magen, "Relations Between Teachers' Subject Matter Knowledge, Teaching Experience and Their Mental Models of Children's Minds and Learning," *Teaching and Teacher Education* 14, no. 6 (1998): 579–95, doi:10.1016/S0742-051X(98)00009-2; Sidney Strauss, Dorit Ravid, Hanna Zelcer, and David C. Berliner, "Teachers' Subject Matter Knowledge and Their Belief Systems About Children's Learning," in *Learning to Read: An Integrated View from Research and Practice*, ed. Terezinha Nunes (London: Kluwer, 1999), 259–82; Sidney Strauss, "Folk Psychology About Others' Learning," in *Encyclopedia of the Sciences of Learning*, 3rd ed., ed. Norbert M. Seel (Heidelberg: Springer, 2011), 1310–13, doi:10.1007/978-1-4419-1428-6.

3. Understanding the Learning Brain

1. Kurt W. Fischer and Juliana Paré-Blagoev, "From Individual Differences to Dynamic Pathways of Development," *Child Development* 71, no. 4 (2000): 850–53, doi:10.1111/1467-8624.00188; Michael J. Karcher and Kurt W. Fischer, "A Developmental Sequence of

Skills in Adolescents' Intergroup Understanding," *Journal of Applied Developmental Psychology* 25, no. 3 (2004): 259–82, doi:10.1016/j.appdev.2004.04.001; Mel Levine, *A Mind at a Time* (New York: Simon & Schuster, 2002).

2. Esther Thelen and Linda B. Smith, "A Dynamic Systems Approach to the Development of Cognition and Action," *Journal of Cognitive Neuroscience* 7, no. 4 (1995): 512–14, doi:10.1162/jocn.1995.7.4.512.

3. Kurt W. Fischer and Samuel P. Rose, "Growth Cycles of Brain and Mind," *Educational Leadership* 56, no. 3 (1998): 56–60; Kurt W. Fischer and Thomas R. Bidell, "Dynamic Development of Action and Thought," in *Handbook of Child Psychology*, 6th ed., ed. William Damon and Richard M. Lerner (Hoboken, NJ: John Wiley & Sons, 2006), 313–99.

4. Marc S. Schwartz and Kurt W. Fischer, "Useful Metaphors for Tackling Problems in Teaching and Learning," *About Campus* 11, no. 1 (2006): 4, doi:10.1002/abc.154.

5. National Scientific Council on the Developing Child, *Excessive Stress Disrupts the Architecture of the Developing Brain: Working Paper 3*, Updated Edition (2005/2014), retrieved from www.developingchild.harvard.edu; Jack P. Shonkoff, "Building a New Biodevelopmental Framework to Guide the Future of Early Childhood Policy," *Child Development* 81, no. 1 (2010): 357–67, doi:10.1111/j.1467-8624.2009.01399.x.

6. R. Keith Sawyer, *The Cambridge Handbook of the Learning Sciences* (New York: Cambridge University Press, 2006).

7. Antonio Damasio, *Descartes' Error: Emotion, Reason, and the Human Brain* (New York: Penguin, 2005); Mary Helen Immordino-Yang and Antonio Damasio, "We Feel, Therefore We Learn: The Relevance of Affective and Social Neuroscience to Education," *Mind, Brain, and Education* 1, no. 1 (2007): 3–10, doi:10.1111/j.1751-228X.2007.00004.x; Mary Helen Immordino-Yang, "The Smoke Around Mirror Neurons: Goals as Sociocultural and Emotional Organizers of Perception and Action in Learning," *Mind, Brain, and Education* 2, no. 2 (2008): 67–73, doi:10.1111/j.1751-228X.2008.00034.x.

8. Immordino-Yang and Damasio, "We Feel, Therefore We Learn," 7.

9. Damasio, *Descartes' Error*; Immordino-Yang, "Smoke Around Mirror Neurons"; Joseph E. LeDoux, "Emotion: Clues from the Brain," *Annual Review of Psychology* 46 (1995): 209–35, doi:10.1146/annurev.ps.46.020195.001233.

10. Damasio, *Descartes' Error*; Michael F. Mascalo and Kurt W. Fischer, "The Dynamic Development of Thinking, Feeling and Acting Over the Life Span," in *The Handbook of Life-Span Development: Cognition, Biology, and Methods*, ed. Richard M. Lerner and Willis F. Overton (Hoboken, NJ: Wiley, 2010), 1:149–94.

4. Why Do We Need Teachers?

1. Vanessa Rodriguez, "The Potential of Systems Thinking in Teacher Reform as Theorized for the Teaching Brain Framework," *Mind, Brain, and Education* 7, no. 2 (2013): 77–85, doi:10.1111/mbe.12013.

2. Alexis Kent, "Synchronization as a Classroom Dynamic: A Practitioner's Perspective," *Mind, Brain, and Education* 7, no. 1 (2013): 13–18, doi:10.1111/mbe.12002; Vanessa Rodriguez, "The Teaching Brain and the End of the Empty Vessel," *Mind, Brain, and Education* 6, no. 4 (2012): 177–85, doi:10.1111/j.1751-228X.2012.01155.x; Katsumi Watanabe, "Teaching as a Dynamic Phenomenon with Interpersonal Interactions," *Mind, Brain, and Education* 7, no. 2 (2013): 91–100, doi:10.1111/mbe.12011; Kazuo Yano, "The Science of Human Interaction and Teaching," *Mind, Brain, and Education* 7, no. 1 (2013): 19–29, doi:10.1111/mbe.12003.

3. Peter Checkland, *Systems Thinking, Systems Practice* (Chichester, UK: Wiley, 1981).

4. Kent, "Synchronization as a Classroom Dynamic."

5. Katherine Clunis D'Andrea, "Trust: A Master Teacher's Perspective on Why It Is Important: How to Build It and Its Implications for MBE Research," *Mind, Brain, and Education* 7, no. 2 (2013): 86–90, doi:10.1111/mbe.12010.

5. Teaching as a System of Skills

1. Mark T. Greenberg, Celene Domitrovich, and Brian Bumbarger, "The Prevention of Mental Disorders in School-Aged Children: Current State of the Field," *Prevention and Treatment* 4, no. 1 (2001), doi:10.1037/1522-3736.4.1.41a; Bridget K. Hamre and Robert C. Pianta, "Learning Opportunities in Preschool and Early Elementary Classrooms," in *School Readiness and the Transition to Kindergarten in the Era of Accountability*, ed. Robert C. Pianta, Martha J. Cox, and Kyle LaBrie Snow (Baltimore: Paul H. Brookes, 2007), 49–83; Frederick J. Morrison and Carol McDonald Connor, "Understanding Schooling

Effects on Early Literacy: A Working Research Strategy," *Journal of School Psychology* 40, no. 6 (2002): 493–500, doi:10.1016/S0022-4405(02)00127-9; Robert C. Pianta, "Teacher-Child Relationships and Early Literacy," in *Handbook of Early Literacy Research*, ed. Susan B. Neuman and David K. Dickinson (New York: Guilford Press, 2006), 2:149–62; Michael Rutter and Barbara Maughan, "School Effectiveness Findings 1979–2002," *Journal of School Psychology* 40, no. 6 (2002): 451–75, doi:10.1016/S0022-4405(02)00124-3.

2. Council for Exceptional Children, *Universal Design for Learning: A Guide for Teachers and Education Professionals.* (Upper Saddle River, NJ: Prentice Hall, 2005); Tracey E. Hall, Anne Meyer, and David H. Rose, *Universal Design for Learning in the Classroom: Practical Applications* (New York: Guilford Press, 2012).

3. Kurt W. Fischer and Thomas R. Bidell, "Dynamic Development of Action and Thought," in *Handbook of Child Psychology*, 6th ed., ed. William Damon and Richard M. Lerner (Hoboken, NJ: John Wiley & Sons, 2006), 313–99; Eric P. Jensen, *Brain-Based Learning: The New Paradigm of Teaching*, 2nd ed. (Thousand Oaks, CA: Corwin Press, 2008); Daniel T. Willingham, *Why Don't Students Like School?: A Cognitive Scientist Answers Questions About How the Mind Works and What It Means for Your Classroom* (San Francisco: Jossey-Bass, 2009).

6. Becoming an Expert Teacher

1. Vanessa Rodriguez and S. Lynneth Solis, "Teachers' Awareness of the Learner–Teacher Interaction: Preliminary Communication of a Study Investigating the Teaching Brain," *Mind, Brain, and Education* 7, no. 3 (2013): 161–69, doi:10.1111/mbe.12023.

2. Vanessa Rodriguez, "The Human Nervous System: A Framework for Teaching and the Teaching Brain," *Mind, Brain, and Education* 7, no. 1 (2013): 2–12, doi:10.1111/mbe.12000

3. Fredric H. Jones with Patrick Jones and Jo Lynne Jones, *Tools for Teaching: Discipline, Instruction, Motivation*, 1st ed. (Santa Cruz, CA: F.H. Jones & Associates, 2000); Diane Ravitch, *The Death and Life of the Great American School System: How Testing and Choice Are Undermining Education* (New York: Basic Books, 2010); Michelle Rhee, *Radical: Fighting to Put Students First* (New York: HarperCollins, 2013).

4. Tovah P. Klein, *How Toddlers Thrive: What Parents Can Do Today*

for Children Ages 2–5 to Plant the Seeds of Lifelong Success (New York: Touchstone, 2014);

5. Susan Moore Johnson, "Incentives for Teachers: What Motivates, What Matters," *Educational Administration Quarterly* 22, no. 3 (1986): 54–79, doi:10.1177/0013161X86022003003; Susan Moore Johnson and the Project on the Next Generation of Teachers, *Finders and Keepers: Helping New Teachers Survive and Thrive in Our Schools*, 1st ed. (San Francisco: Jossey-Bass, 2004); Kavita Kapadia and Vanessa Coca with John Q. Easton, *Keeping New Teachers: A First Look at the Influences of Induction in the Chicago Public Schools* (Chicago: Chicago Consortium for School Research, 2007); Jessica Levin, Jennifer Mulhern, and Joan Schunck, *Unintended Consequences: The Case for Reforming the Staffing Rules in Urban Teachers Union Contracts* (New York: New Teacher Project, 2005).

6. Christopher Clark and Magdalene Lampert, "The Study of Teacher Thinking: Implications for Teacher Education," *Journal of Teacher Education* 37, no. 5 (1986): 27–31, doi:10.1177/002248718603700506.

7. Lorenzo Cherubini, "A Grounded Theory of Prospective Teachers' Meta-cognitive Process: Internalizing the Professional Standards of Teaching," *Teacher Educator* 45, no. 2 (2010): 96–117, doi:10.1080/08878731003628593; David Premack and Guy Woodruff, "Does the Chimpanzee Have a Theory of Mind?," *Behavioral and Brain Sciences* 1, no. 4 (1978): 515–526, doi:10.1017/S0140525X00076512.

8. Bridget K. Hamre and Robert C. Pianta, "Learning Opportunities in Preschool and Early Elementary Classrooms," in *School Readiness and the Transition to Kindergarten in the Era of Accountability*, ed. Robert C. Pianta, Martha J. Cox, and Kyle LaBrie Snow (Baltimore: Paul H. Brookes, 2007), 49–83; Karen M. La Paro, Robert C. Pianta and Megan Stuhlman, "The Classroom Assessment Scoring System: Findings from the Prekindergarten Year," *Elementary School Journal* 104, no. 5 (2004): 409–26.

9. Gail Boushey and Joan Moser, *The Daily 5: Fostering Literacy Independence in the Elementary Grades*, 1st ed. (Portland: Stenhouse, 2006); Jones et al., *Tools for Teaching*; Doug Lemov, *Teach Like a Champion: 49 Techniques That Put Students on the Path to College*, 1st ed. (San Francisco: Jossey-Bass, 2010).

10. Nancie Atwell, *In the Middle: New Understanding About Writing, Reading, and Learning*, 2nd ed. (Portsmouth, NH: Boynton/Cook,

1998); Lucy Calkins, Mary Ehrenworth, and Christopher Lehman, *Pathways to the Common Core: Accelerating Achievement* (Portsmouth, NH: Heinemann, 2012); Charlotte Danielson, *The Handbook for Enhancing Professional Practice: Using the Framework for Teaching in Your School* (Alexandria, VA: Association for Supervision and Curriculum Development, 2008).

11. Sanjoy Mahajan, "To Develop Expertise, Motivation Is Necessary But Insufficient," Freakonomics.com, November 25, 2011, http://freakonomics.com/2011/11/25/to-develop-expertise-motivation-is -necessary-but-insufficient.

12. Atwell, *In the Middle*; Calkins et al., *Pathways to the Common Core*; Howard Gardner, *Frames of Mind: The Theory of Multiple Intelligences* (New York: Basic Books, 1983).

13. Danielson, *Handbook for Enhancing Professional Practice*; Mary Ehrenworth, *Looking to Write: Children Writing Through the Visual Arts* (Portsmouth, NH: Heinemann, 2003).

14. Charlotte Danielson, *Teaching Methods* (Upper Saddle River, NJ: Merrill/Pearson, 2010).

15. Ibid.; Grant Wiggins and Jay McTighe, *Understanding by Design*, 2nd ed. (Alexandria, VA: Association for Supervision and Curriculum Development, 2005).

16. Lemov, *Teach Like a Champion*; Jay Mathews, *Work Hard. Be Nice.: How Two Inspired Teachers Created the Most Promising Schools in America*, 1st ed. (Chapel Hill, NC: Algonquin Books of Chapel Hill, 2009); Paul Tough, *Whatever It Takes: Geoffrey Canada's Quest to Change Harlem and America* (Boston: Houghton Mifflin, 2008).

17. Daniel H. Pink, *Drive: The Surprising Truth About What Motivates Us* (New York: Riverhead Books, 2009).

18. Clark and Lampert, "Study of Teacher Thinking."

19. Mihaly Csikszentmihalyi, *Flow: The Psychology of Optimal Experience* (New York: HarperPerennial, 1991); Marleen B. Schippers, Alard Roebroeck, Remco Renken, Luca Nanetti, and Christian Keysers, "Mapping the Information Flow from One Brain to Another During Gestural Communication," *Proceedings of the National Academy of Sciences* 107, no. 20 (2010): 9388–93, doi:10.1073/pnas.1001791107; Kazuo Yano, "The Science of Human Interaction and Teaching," *Mind, Brain, and Education* 7, no. 1 (2013): 19–29, doi:10.1111/mbe.12003.

20. Yano, "Science of Human Interaction"; Alexis Kent, "Syn-

chronization as a Classroom Dynamic: A Practitioner's Perspective," *Mind, Brain, and Education* 7, no. 1 (2013): 13–18, doi:10.1111 /mbe.12002.

21. Csikszentmihalyi, *Flow*.

7. Developing the Teaching Brain

1. Kurt W. Fischer and Thomas R. Bidell, "Dynamic Development of Action and Thought," in *Handbook of Child Psychology*, 6th ed., ed. William Damon and Richard M. Lerner (Hoboken, NJ: John Wiley & Sons, 2006), 313–99

2. Gergely Csibra and György Gergely, "Natural Pedagogy as Evolutionary Adaptation," *Philosophical Transactions of the Royal Society B: Biological Sciences* 366, no. 1567 (2011): 1149–57, doi:10.1098/ rstb.2010.0319.

3. Thomas S. Kuhn, *The Structure of Scientific Revolutions*, 2nd ed. (Chicago: University of Chicago Press, 1970).

4. Sidney Strauss and Margalit Ziv, "Teaching Is a Natural Cognitive Ability for Humans," *Mind, Brain, and Education* 6, no. 4 (2012): 186– 96, doi:10.1111/j.1751-228X.2012.01156.x

5. Charlotte Danielson, *Teaching Methods* (Upper Saddle River, NJ: Merrill/Pearson, 2010); Grant Wiggins and Jay McTighe, *Understanding by Design*, 2nd ed. (Alexandria, VA: Association for Supervision and Curriculum Development, 2005).

8. *Your* Teaching Brain

1. Carl D. Marci and Helen Riess, "Physiologic Monitoring in Psychodynamic Psychotherapy Research," in *Handbook of Evidence-Based Psychodynamic Psychotherapy: Bridging the Gap Between Science and Practice*, ed. Raymond A. Levy and J. Stuart Ablon (New York: Humana Press, 2008), doi:10.1007/978-1-59745-444-5_14.

9. The Teaching Brain and Next Steps for Education Reform

1. Lisa D. Delpit, "The Silenced Dialogue: Power and Pedagogy in Educating Other People's Children," *Harvard Educational Review* 58, no. 3 (1988): 280–99; Lisa Delpit, "Seeing Color: A Review of *White Teacher*," in *Rethinking Our Classrooms: Teaching for Equity and Justice*, 2nd ed., ed. Wayne Au, Bill Bigelow, and Stan Karp (Milwaukee, WI: Rethinking Schools, 2007), 158–60.

2. Lisa Delpit and Theresa Perry, eds., *The Real Ebonics Debate: Power, Language, and the Education of African-American Children* (Boston, MA: Beacon Press, 1998); Lisa Delpit and Joanne Kilgour Dowdy, eds., *The Skin That We Speak: Thoughts on Language and Culture in the Classroom* (New York: The New Press, 2002).

3. L. Todd Rose with Katherine Ellison, *Square Peg: My Story and What It Means for Raising Innovators, Visionaries, and Out-of-the-Box Thinkers* (New York: Hyperion, 2013).

4. Kyongsik Yun, Katsumi Watanabe, and Shinsuke Shimojo, "Interpersonal Body and Neural Synchronization as a Marker of Implicit Social Interaction," *Scientific Reports* 2, no. 959 (2011), doi:10.1038/srep00959.

5. Lisa Holper, Andrea P. Goldin, Diego E. Shalóm, Antonio M. Battro, Martin Wolf, and Mariano Sigman, "The Teaching and the Learning Brain: A Cortical Hemodynamic Marker of Teacher-Student Interactions in the Socratic Dialog," *International Journal of Educational Research* 59 (2013): 1–10, doi:10.1016/j.ijer.2013.02.002.

6. José-Rodrigo Córdoba-Pachón, "Embracing Human Experience in Applied Systems-Thinking," *Systems Research and Behavioral Science* 28, no. 6 (2011): 680–88, doi:10.1002/sres.1117.

Appendix B: Methods Used to Investigate the Cognitive Processes of the Teaching Brain

1. Robert S. Siegler and Kevin Crowley, "The Microgenetic Method: A Direct Means for Studying Cognitive Development," *American Psychologist* 46 (1991): 606–20; Nira Granott and Jim Parziale, "Microdevelopment: A Process-Oriented Perspective for Studying Development and Learning," in *Microdevelopment: Transition Processes in Development and Learning*, ed. Nira Granott and Jim Parziale (New York: Cambridge University Press, 2002), 1–28; Deanna Kuhn, "Microgenetic Study of Change: What Has It Told Us?," *Psychological Science* 6, no. 3 (1995): 133–39, doi:10.1111/j.1467-9280.1995.tb00322.x.

2. Ralph T. Putnam and Hilda Borko, "What Do New Views of Knowledge and Thinking Have to Say about Research on Teacher Learning?," *Educational Researcher* 29, no. 1 (2000): 4–15.

3. Siegler and Crowley, "Microgenetic Method."

4. Stephen Bochner, "Cross-Cultural Differences in the Self Concept: A Test of Hofstede's Individualism/Collectivism Distinction,"

Journal of Cross-Cultural Psychology 25, no. 2 (1994): 273–83, doi:10 .1177/0022022194252007; Steven D. Cousins, "Culture and Self-Perception in Japan and the United States," *Journal of Personality and Social Psychology* 56, no. 1 (1989): 124–31.

5. Kurt W. Fischer, "A Theory of Cognitive Development: The Control and Construction of Hierarchies of Skills," *Psychological Review* 87, no. 6 (1980): 477–531.

6. C.L. Cheng, "Constructing Self-Representations Through Social Comparison in Peer Relations: The Development of Taiwanese Grade-School Children," 1999, Dissertation Abstracts International: Section B: The Sciences and Engineering, 59(9-B), 5132; Kurt W. Fischer, "Dynamic Cycles of Cognitive and Brain Development: Measuring Growth in Mind, Brain, and Education," in *The Educated Brain: Essays in Neuroeducation*, ed. Antonio M. Battro, Kurt W. Fischer, and Pierre J. Léna (Cambridge: Cambridge University Press, 2008), 127–50.

7. Vanessa Rodriguez, "The Teaching Brain and the End of the Empty Vessel," *Mind, Brain, and Education* 6, no. 4 (2012): 177–85, doi:10.1111/j.1751-228X.2012.01155.x.

8. Lee S. Shulman, "Knowledge and Teaching: Foundations of the New Reform," *Harvard Educational Review* 57, no. 1 (1987): 1–23.

9. Gaea Leinhardt and James G. Greeno, "The Cognitive Skill of Teaching," *Journal of Educational Psychology* 78, no. 2 (1986): 75–95, doi:10.1037/0022-0663.78.2.75; Penelope L. Peterson and Christopher M. Clark, "Teachers' Reports of Their Cognitive Processes During Teaching," *American Educational Research Journal* 15, no. 4 (1978): 555–65.

10. Kate Eliza O'Connor, "You Choose to Care: Teachers, Emotions and Professional Identity," *Teaching and Teacher Education* 24, no. 1 (2008): 117–26; Alfredo Urzúa and Vásquez, "Reflection and Professional Identity in Teachers' Future-Oriented Discourse," *Teaching and Teacher Education* 24, no. 7 (2008): 1935–46.

11. Ryuta Aoki, Tsukasa Funane, and Hideaki Koizumi, "Brain Science of Ethics: Present Status and the Future," *Mind, Brain, and Education* 4, no. 4 (2010): 188–95, doi:10.1111/j.1751-228X.2010.01098.x; Battro et al., *Educated Brain*; Antonio M. Battro, "The Teaching Brain," *Mind, Brain, and Education* 4, no. 1 (2010): 28–33, doi:10.1111/j.1751-228X.2009.01080.x; Richard E. Passingham, *What Is Special About the Human Brain?* (New York: Oxford University Press, 2008).

12. Kurt W. Fischer, Todd Rose, and Samuel P. Rose, "Growth Cycles of Mind and Brain: Analyzing Developmental Pathways of Learning Disorders," in *Mind, Brain, and Education in Reading Disorders*, ed. Kurt W. Fischer, Jane Holmes Bernstein, and Mary-Helen Immordino-Yang (New York: Cambridge University Press, 2007), 101; David Premack and Ann James Premack, "Why Animals Lack Pedagogy and Some Cultures Have More of It Than Others," in *The Handbook of Education and Human Development: New Models of Learning, Teaching and Schooling*, ed. David R. Olson and Nancy Torrance (Cambridge, MA: Blackwell, 1996), 302–44; Sidney Strauss, "Teaching as a Natural Cognitive Ability: Implications for Classroom Practice and Teacher Education," in *Developmental Psychology and Social Change: Research, History, and Policy*, ed. David B. Pillemer and Sheldon H. White (New York: Cambridge University Press, 2005), 368–89.

13. T.M. Caro and M.D. Hauser, "Is There Teaching in Nonhuman Animals?," *Quarterly Review of Biology* 67, no. 2 (1992): 151–74; Alex Thornton and Nichola J. Raihani, "The Evolution of Teaching," *Animal Behaviour* 75, no. 6 (2008): 1823–36, doi:10.1016/j.anbehav.2007.12.014.

14. Christopher M. Clark and Penelope L. Peterson, "Teachers' Thought Processes," in *Handbook of Research on Teaching: A Project of the American Educational Research Association*, 3rd ed., ed., Merlin C. Wittrock (New York: Macmillan, 1986), 255–314; Suzanne M. Wilson, Lee S. Shulman, and Anna E. Richert, "'150 Ways of Knowing': Representations of Knowledge in Teaching," in *Exploring Teachers' Thinking*, ed. James Calderhead (London: Cassell, 1984), 104–24.

Appendix C: Types of Teaching Ability Developed Across the Life Span

1. Adapted from Sidney Strauss and Margalit Ziv, "Teaching Is a Natural Cognitive Ability for Humans," *Mind, Brain, and Education* 6, no. 4 (2012): 186–96, doi:10.1111/j.1751-228X.2012.01156.x.

2. Kazushige Akagi, "Development of Teaching Behavior in Typically Developing Children and Children with Autism," in *CARLS Series of Advanced Study of Logic and Sensibility* (Tokyo: Keio University Press, 2012), 5:425–35.

3. Sidney Strauss, Margalit Ziv, and Adi Stein, "Teaching as a Natural Cognition and Its Relations to Preschoolers' Developing Theory of Mind," *Cognitive Development* 17, no. 3–4 (2002): 1473–87,

doi:10.1016/S0885-2014(02)00128-4; Margalit Ziv and Douglas Frye, "Children's Understanding of Teaching: The Role of Knowledge and Belief," *Cognitive Development* 19, no. 4 (2004): 457–77; Margalit Ziv, Ayelet Solomon, and Douglas Frye, "Young Children's Recognition of the Intentionality of Teaching," *Child Development* 79, no. 5 (2008): 1237–56, doi:10.1111/j.1467-8624.2008.01186.x.

4. Leila Bensalah, Marie Olivier, and Nicolas Stefaniak, "Acquisition of the Concept of Teaching and Its Relationship with Theory of Mind in French 3- to 6-Year-Olds," *Teaching and Teacher Education* 28, no. 3 (2012): 303–11; Strauss et al., "Teaching as a Natural Cognition"; Ziv and Frye, "Children's Understanding of Teaching"; Ziv et al., "Young Children's Recognition."

5. David Wood, Heather Wood, and David Middleton, "An Experimental Evaluation of Four Face-to-Face Teaching Strategies," *International Journal of Behavioral Development* 1, no. 2 (1978): 131–47, doi:10.1177/016502547800100203.

6. James Garbarino, "The Impact of Anticipated Reward Upon Cross-Age Tutoring," *Journal of Personality and Social Psychology* 32, no. 3 (1975): 421–28; Russell J. Ludeke and Willard W. Hartup, "Teaching Behaviors of 9- and 11-Year-Old Girls in Mixed-Age and Same-Age Dyads," *Journal of Educational Psychology* 75, no. 6 (1983): 908–14, doi:10.1037/0022-0663.75.6.908.

7. Sidney Strauss, "Theories of Learning and Development for Academics and Educators," *Educational Psychologist* 28, no. 3 (1993): doi:10.1207/s15326985ep2803_1; Sidney Strauss, "Folk Psychology, Folk Pedagogy and Their Relations to Subject Matter Knowledge," in *Understanding and Teaching the Intuitive Mind*, ed. Bruce Torff and Robert J. Sternberg (Mahwah, NJ: Erlbaum, 2001), 217–42; Sidney Strauss, "Folk Psychology About Others' Learning," in *Encyclopedia of the Sciences of Learning*, 3rd ed., ed. Norbert M. Seel (Heidelberg: Springer, 2011), 1310–13, doi:10.1007/978-1-4419-1428-6; Sidney Strauss and Tamar Shilony, "18 Teachers' Models of Children's Minds and Learning," in *Mapping the Mind: Domain Specificity in Cognition and Culture*, ed. Lawrence A. Hirschfeld and Susan A. Gelman (Cambridge: Cambridge University Press, 1994), 455–73.

INDEX

Note: Page numbers in bold italics represent figures.

animal-behaviorist models: animal learning studies, 5–8, 10, 30, 32–33, 34–35, 86–88; and differences between human and animal learning, 54–55; education reforms built on, 32–33, 87–88; fast-track teacher training programs, 32–33, 87–88; Pavlov's dogs, 5, 30; pied babblers, 6–7, 30, 86–87; and spinal cord teaching, 86–87. *See also* behaviorist views of learning/teaching

author's teaching experiences, xii–xiii, 67–69; ELA teaching, 27–30, 93–96; as experienced teacher, 43–46, 91–97; and NCLB mandated exams, 27–30; as novice teacher, 3–5, 44–45; and shy students, 176; and student with a "different" learning style, 43–46; student-centered teaching (the case of Logan), 91–97

awareness of context (AoC), **99**, 121–23, **124**, **131**, 150, 151, **153**, 154; and AoL, 121–22; and AoST, 121–22; deep reflection questions to help teachers develop, 167–68; and external influences on the teacher, 122–23, 167–68; limitations in children teachers, 142; scenario of middle school student and expert teacher, 150, 151, **153**, 154

awareness of learner (AoL), **99**, 104–8, 123, **124**, **131**, 137, 142, 144–46, **147**, 149, 150, 151, 152, **153**; and AoC, 121–22; deep reflection questions to help teachers develop, 162–63; forming a theory of cognition, 104–5, 123, 149, 150, 151, **153**, 162; forming a theory of emotion, 106–8, 123, 149, **153**, 163; forming a theory of memory, 105–6, 123, 150, **153**, 163; forming a theory of mind (ToM) for the learner, 104, 123, 162;

ABOUT THE AUTHORS

Vanessa Rodriguez taught middle school humanities in the New York City public schools for more than ten years before deciding to return to graduate school in pursuit of understanding better exactly what—beyond her love of children—inspires her love of teaching. She is now a doctoral candidate at the Harvard Graduate School of Education, where her research has been recognized for its innovation and potential impact on education with the prestigious Hauser Initiative for Learning and Teaching (HILT) award. She has a BA in English literature from New York University, an MS in education from the City College of New York, and an EdM in Mind, Brain, and Education from the Harvard Graduate School of Education. A New Yorker at heart, she currently lives in Cambridge, Massachusetts. This is her first book.

Michelle Fitzpatrick has co-authored many nonfiction books, including a number of *New York Times* and international best-sellers. She writes on a wide variety of topics, including health, psychology, sexuality, neuroscience, and education. She has an MA in literature and creative writing from the University of Houston and an EdM in Mind, Brain, and Education from Harvard.

PUBLISHING IN
THE PUBLIC INTEREST

Thank you for reading this book published by The New Press. The New Press is a nonprofit, public interest publisher. New Press books and authors play a crucial role in sparking conversations about the key political and social issues of our day.

We hope you enjoyed this book and that you will stay in touch with The New Press. Here are a few ways to stay up to date with our books, events, and the issues we cover:

- Sign up at www.thenewpress.com/subscribe to receive updates on New Press authors and issues and to be notified about local events

- Like us on Facebook: www.facebook.com/newpressbooks

- Follow us on Twitter: www.twitter.com/thenewpress

Please consider buying New Press books for yourself; for friends and family; or to donate to schools, libraries, community centers, prison libraries, and other organizations involved with the issues our authors write about.

The New Press is a 501(c)(3) nonprofit organization. You can also support our work with a tax-deductible gift by visiting www.thenewpress.com/donate.